Farm Machinery

6TH EDITION

BOOKS AND DVDS BY BRIAN BELL

Books
Farm Machinery 6th edition
Machinery for Horticulture (with Stewart Cousins)
Ransomes, Sims & Jefferies
Seventy Years of Farm Machinery Vol. 1 Seedtime
Seventy Years of Farm Machinery Vol. 2 Harvest
Seventy Years of Farm Tractors
Seventy Years of Garden Machinery
The Tractor Ploughing Manual

DVDs
Acres of Change
Classic Combines
Classic Farm Machinery Vol. 1
Classic Farm Machinery Vol. 2
Classic Tractors
Farm Machinery Film Records Vol. 1
Farm Machinery Film Records Vol. 2
Farm Machinery Film Records Vol. 3
Harvest from Sickle to Satellite
Ploughs and Ploughing Techniques
Power of the Past
Reversible and Conventional Ploughing Skills
Steam at Strumpshaw
Thatcher's Harvest
Tracks Across the Field
Vintage Match Ploughing Skills
Vintage Garden Tractors

Farm Machinery

6TH EDITION

Brian Bell MBE
Martin Rickatson

First published 1979

Second edition 1983

Third edition 1989

Fourth edition 1996

Fifth edition 2005, revised 2008

Published by
5M Publishing Ltd,
Benchmark House,
8 Smithy Wood Drive,
Sheffield, S35 1QN, UK
Tel: +44 (0) 1234 81 81 80
www.5mpublishing.com

A catalogue record for this book is available from the British Library

ISBN 978–1–910456–06–4

Book layout by Apex CoVantage, LLC
Printed by Replika Press Pvt Ltd, India
Photos by Brian Bell unless otherwise credited

CONTENTS

PREFACE TO SIXTH EDITION

The continuing march of farm mechanisation has seen the average engine power of new tractors sold to farmers in the UK rise from approximately 60 kW (80 hp) in the mid-1990s to almost twice that level in 2015. Electronic controls and sophisticated monitoring systems are standard equipment in many tractor cabs, which along with features such as climate control makes them at least as comfortable as the average family car. Automatic steering, GPS mapping of crop yields aided by satellite navigation systems, electronic engine and transmission management and even robotic milking parlours are now part of today's farming scene.

These developments have led to the addition of new material to this sixth edition of *Farm Machinery*. Together with two hundred and eighty new colour plates and one hundred and forty-five line drawings, the revision takes full account of the continuing progress of mechanisation on British farms. However, this has not been done at the expense of omitting earlier material concerning the working principles, and operation of, the vast array of less sophisticated tractors and farm machines still in use on British farms in the twenty-first century.

The new edition includes fully revised chapters on tractors, cultivation and drilling equipment, crop care and harvest machinery. Further chapters deal with farmyard and estate maintenance machinery, mechanical handling, dairy equipment, farm power and the farm workshop. Electronic control and machinery management systems are considered throughout the book. References to the Health and Safety at Work Act and other safety regulations summarise their main requirements, but they should only be taken as a guide. The reader should study the relevant information leaflets on all aspects of farm safety, and in particular those concerning the safe use of pesticides, chain saws and forklifts.

Dimensions and settings are given in metric units, usually with an approximate imperial equivalent.

ACKNOWLEDGEMENTS

I am very grateful to Martin Rickatson, who has edited the manuscript for the new sixth edition of *Farm Machinery* and suggested numerous technical additions, which have been included in the text. He has also sourced a number of colour photographs for the book, which is the first edition to be illustrated in full colour.

Prior to taking a degree in farm management at Wye College, Martin worked on his family's pig, dairy and arable farms before spending two years as arable and machinery writer with *Farming News* and seven years as deputy editor of the farm machinery monthly *Profi International*. He now works as a freelance writer for farming magazines and farm machinery companies, and has contributed to books ranging from tractor encyclopaedias to histories of International Harvester tractors and combines.

We are both grateful to the many individuals and companies involved in the manufacture of tractors and farm machinery who have helped with the preparation of this new edition of *Farm Machinery*.

Our thanks are especially due to Richard Abbott, Michael Alsop, Paul Bellas, Mike Bennett, Simon Brown, Demien Buckle, Lynn Chilvers, Tim Coleridge, Rob Immink, Tom Miller, Steve Mitchell, Andrew Starbuck and Tom Teagle. Many companies, including those listed here, have provided photographs for the new edition and this is acknowledged in the captions. They include Amazone, Bomford Turner, Claas UK, Claydon Yield-O-Meter, Garford Farm Machinery and Kverneland Group. Photographs have also been provided by Landquip Crop Sprayers, Lemken UK, Massey Ferguson, New Holland, Opico, David Ritchie Implements, Shelbourne Reynolds, Suzuki GB, Teagle Machinery, Twose, Vaderstad and Vicon.

CHAPTER 1

Farm Tractors

Tractors have been used on British farms since the early part of the twentieth century. They were already replacing horses on some farms by 1920, when tractors were used to pull implements from the drawbar and drive stationary equipment, such as a threshing machine, by means of a belt pulley.

By the early 1940s, tractor design had progressed to the stage where a hydraulic lift system had been developed, as had the power

Plate 1.1   Powershift transmission, front axle suspension, category II or III hydraulic linkage, front power take-off and a suspended cab are some of the features of this 170 hp tractor. (Massey Ferguson)

take-off shaft used to drive trailed implements such as a self-binder or a manure spreader. During the immediate post-war years, the tractor population on British farms exploded. There were numerous improvements in design: diesel engines, the differential lock, four-wheel drive and cabs to protect the driver from the weather are just a few examples.

Today, many tractors used on arable farms have in-cab electronic controls which allow the driver to carry out various operations from the comfort of an air-conditioned safety cab, and some have front- and rear-mounted cameras to warn the driver of any unseen hazards. Other less sophisticated models with mainly manual controls are used on stock and mixed farms, where higher specifications are not required and their cost cannot be justified.

Electronically controlled hydraulic systems with sensors linked to the control unit, which raise and lower mounted implements to maintain constant draft, are often found on tractors which also feature automatic gear changing systems to achieve optimum performance and economy. Some tractors have a headland management system that automatically carries out the complete sequence of lifting the implement, disengaging the power take-off – when it is being used – and switching off four-wheel drive. After the driver has turned the tractor at the headland turn, the management system reverses the process and returns the implement to its working position. Some management systems also turn the tractor at the headland.

For many years farm tractors had engines of no more than 15–45 kW (20–60 hp), but engine power has steadily increased and the average engine size of tractors now sold to British farmers is in excess of 100 kW (135 hp). Four-wheel drive tractors with engines in the 110–220 kW (150–300 hp) range or more are used on large-scale arable farms. Tractors used on predominately livestock farms are usually less powerful, typically with an engine in the 45–75 kW (60–100 hp) range.

In the early days of farm mechanisation, tractors were used to pull an implement from a fixed or swinging drawbar. The introduction of the hydraulic system in the 1940s brought the versatility we expect of the modern farm tractor, with the ability to lift implements, operate external hydraulic rams and drive machines with hydraulic motors. As well as providing power for everything from the starter motor to in-cab electronic control systems, the tractor electrical system can be used to make adjustments to implements with solenoids and small electric motors.

TYPES OF TRACTOR

Wheeled tractors range from small two-wheel drive models for market gardens to huge four-wheel drive and rubber-tracked machines with

Plate 1.2   Two-wheel drive tractors were in almost universal use until the introduction of four-wheel drive conversion kits in the mid-1960s, but within twenty years four-wheel drive was available for almost all new tractors. (Case IH)

Plate 1.3   Some small farm tractors with a low noise level have a safety roll bar which can be folded when used in low farm buildings. (John Deere)

engines developing 150–300 kW (200–400 hp) or more.

Rowcrop Tractors

Mainly used by farmers and market gardeners for vegetable and root crop production, rowcrop tractors have a small turning circle, easily adjusted wheel track settings and provide the driver with a full view of the crop when inter-row hoeing and for rowcrop work. Few modern tractors meet these features, as they are too large and powerful for this type of work, but many older models of rowcrop tractor are still in use.

Compact tractors with a diesel engine in the 15–30 kW (20–40 hp) range are ideal for working in rowcrops. A typical model has a three-cylinder diesel engine with a manual or optional hydrostatic transmission, power take-off and hydraulics and manual or power steering. A typical compact tractor with a manual gearbox has twelve forward and four reverse gears. Some rowcrop tractors may have four-wheel drive, a low-speed creeper gearbox and a safety cab or frame. Small, narrow track rowcrop tractors with an overall width of 52 in. or less are used for orchard and vineyard work.

Plate 1.4   General-purpose tractors with 60–150 kW (80–200 hp) engines do much of the general work on arable and livestock farms. (New Holland)

General-Purpose Tractors

As the name suggests, these tractors are used for most of the day-to-day work on arable and livestock farms. Engine power ranges from 60–150 kW (80–200 hp) or more depending on the size and type of tractor. They are capable of pulling heavy loads from the hydraulic linkage, pick-up hitch or drawbar and have high capacity hydraulic systems and powerful power take-off shafts.

Almost all modern general-purpose tractors have four-wheel drive, but small two-wheel drive models are still in widespread use, especially in mixed farming areas. To aid steering, the front wheels on most four-wheel drive tractors are smaller in diameter than the rear wheels, but some older four-wheel drive models have front and rear wheels of equal size or front wheels that are much smaller than those at the rear.

Many of the more powerful general-purpose tractors have front and rear hydraulic linkage, power take-off shafts, disc brakes on all four wheels, diff-locks on both axles and some have a pneumatic or a mechanical suspension system. Some very high power four-wheel drive models with 250–390 kW (340–530 hp) engines, infinitely variable speed transmissions and a top speed of 50 kph (30 mph) have a cab which the driver can rotate from the driving seat so that the tractor can be driven in either direction.

High mobility tractors with similar specifications to general-purpose models may have a

Plate 1.6   Four-wheel steering, suspension on all four wheels and a turbocharged diesel engine are features of this high-speed four-wheel drive tractor.

load carrying platform at the rear, which can be used with a de-mountable crop sprayer or fertiliser spreader. Suitable for both fieldwork and high-speed road haulage, these tractors have transmission systems with top road speeds of 64–80 kph (40–50 mph) when equipped with suitable tyres and an approved braking system.

Tracklayers

Small tracklayers or crawlers with 45–60 kW (60–80 hp) diesel engines and steel tracks are still used on some heavy land farms. Tracklayers have the advantage of a large footprint area and cause less soil compaction than an equivalent-size wheeled tractor. However, most farmers wishing to reduce soil compaction now use a rubber-tracked tracklayer or a four-wheel drive tractor.

Although some large steel-tracked tracklayers with diesel engines in the 67–100 kW (85–150 hp) power range are owned by farmers, most of them are used by contractors for land drainage work. While this type of tractor has rather low operating speeds, the tracks exert very low ground pressure compared with wheels, which means soil compaction is reduced to a minimum. However, steel-tracked models may only be driven on public

Plate 1.5   The cab on this multi-purpose tractor can be turned round to face the required direction of travel. This tractor is pushing a large set of mounted mowers. (Claas)

Plate 1.7   Track plates have to be fitted to crawler tractors with steel tracks before they can be driven short distances on the public highway, a sight now seldom seen.

Plate 1.8   The first crawler tractors with rubber tracks had equal-sized drive sprockets and idler wheels. (Gregoire-Besson)

Plate 1.9   This rubber-tracked crawler with a 16-speed powershift transmission has a top road speed of 30 kph (20 mph). (John Deere)

roads after street plates have been fitted to the tracks. Alternatively, they must be transported between sites on a trailer.

Rubber-tracked crawler tractors can be driven on the public highway. They have top speeds of up to 40 kph (25 mph) and, depending on model, have a six-cylinder engine in the 150–450 kW (200–600 hp) power range. The tracks are driven by sprockets which engage with lugs on the inside of the tracks. Diagonal lugs on the outer surface of the tracks give maximum power transmission. Rubber-tyred track rollers and a front idler wheel support the track. Some rubber-tracked models have a conventional gear-based powershift transmission; others have an infinitely variable hydrostatic drive to each track.

The very large footprint area of both rubber and steel tracks compared with a four-wheel drive tractor gives more efficient use of engine power at the drawbar and improved traction.

Plate 1.10   With a six-cylinder 335 kW (450 hp) engine, this tractor is a wheeled, articulated alternative to the rigid-frame, rubber-tracked crawler shown in Plate 1.11. (Gregoire-Besson)

Plate 1.11   A powershift transmission, four independently driven rubber tracks and articulated centre-pivot steering make this tractor easy to manoeuvre on headlands.

CONTROLS AND INSTRUMENTS

The Driving Controls

Although some tractors have only a basic set of manual instruments and controls, most modern tractors have various switches, buttons, gauges and levers for operation, control and monitoring, and some have an on-board computer which carries out the same functions.

Controls and instruments vary in quantity and complexity depending on the size and cost of the tractor. They are also arranged in the cab in many different ways, and it is important to be familiar with the controls before using a particular model for the first time. The basic driving controls are:

Throttle This sets the engine speed. The hand throttle sets the engine speed for continuous work such as ploughing and the engine governor maintains this speed. The foot throttle is used where frequent changes in speed are needed for such work as manure loading and road transport. The foot throttle will over-ride the hand throttle but when the pedal is released the engine speed returns to the hand throttle setting.

Some tractors have computer-controlled speed matching which automatically increases engine speed when an increased load, such as driving uphill, causes the engine speed to fall away and then, as the load is reduced, the computer brings the engine back to the selected speed.

Clutch This engages and disengages the drive between the engine and transmission. Depending on the type of transmission the clutch may be controlled with a pedal or an automatic clutch pack.

Some older or very basic tractors have a dual clutch; pushing the pedal halfway down stops forward travel, and pushing it fully down disengages the power take-off shaft.

Gear levers are used to select the speed range, forward, reverse and neutral. Older tractors with a manual gearbox usually have two levers, one to select the gear and the second for forward or reverse and the high or low range of gears.

Many tractors have a separate lever or switch to engage a two or three step 'splitter' or powershift, which without using the clutch provides a high and low ratio in each gear. Tractors with a change-on-the-move (semi- or full-powershift) transmission have a lever or switch to select gears and another to select forward or reverse.

Brakes are used to slow, stop or park the tractor and assist with steering, especially when making headland turns. Most tractors have separate pedals for the left- and right-hand brakes and these must be locked together for roadwork. They are usually unlatched in the field so that the inner wheel can be braked to shorten the turning circle when making a headland turn. The parking brake may be a latch used to lock both pedals in the down position or a handbrake on the transmission system. Tractors with a change-on-the-move transmission have a neutral or park position built into the control lever. Most four-wheel drive tractors have hydraulically operated disc brakes on all four wheels.

The differential lock (diff-lock) prevents wheel spin by an individual wheel. Most four-wheel drive tractors have a diff-lock on both axles and when it is engaged, either with a switch or a pedal, both wheels on the same axle can only turn at the same speed. This reduces wheel spin and increases traction in difficult soil conditions. The diff-lock is designed to disengage automatically but can be released by briefly applying either a brake pedal or the clutch pedal. Many tractors have a diff-lock which is engaged and disengaged with a switch.

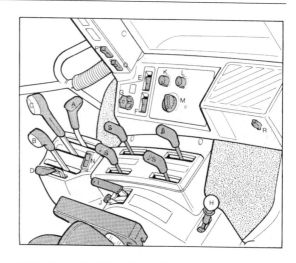

Figure 1.1   Transmission and hydraulic controls.

A – Gear range lever. B – Shuttle lever. C – Main gear lever and powershift control. D – Hand throttle. E – Diff-lock switch. F – Four-wheel drive switch. G – Power take-off switch. H – Power take-off range selector. I – Hydraulic lift lever. J – Adjustable stop for hydraulic lever. K – Drop rate control. L – Height limiting control. M – Position and draft control sensitivity. N – Hydraulic lift raise and lower switch. O – Roof beacon switch. P – External four-pin electric socket switch. R – Power socket for electronic implement monitoring equipment. S – Control levers for auxiliary service valves.

Plate 1.12   The manual driving controls on this tractor include throttle lever (A), individual brake pedals (B), clutch pedal (C) and range and gear levers (D).

Power take-off drive is engaged with a switch or lever-operated hydraulic clutch unit. Older tractors have a lever, used together with the main clutch pedal, to engage the drive to the power take-off.

A two-speed power shaft with speeds of 540 and 1,000 rpm is standard on many tractors. The 1,000 rpm setting can be used at a reduced engine speed to give an economy speed of 540 rpm for light power take-off work.

Hydraulics Depending on the age and model of the tractor, levers, switches or buttons are used to operate the hydraulic system. The basic hydraulic controls are:

1. Draft control is used with ploughs, cultivators and other implements that work in the soil. It maintains an even draft and an almost regular depth without the need for depth wheels,

provided that soil conditions are reasonably constant. Some heavy mounted implements also have depth wheels to aid depth control.

2. Position control is used for mounted implements such as sprayers or hedge cutters that need to be held at a constant height above the ground.

3. Hydraulic couplings at the back of the tractor can be connected with hydraulically-operated farm implements. Each coupling has a separate control lever. They supply single-acting rams which exert force in one direction or double-acting rams which operate in two directions, e.g. up and down or in and out.

4. Remote lift and drop control switches on the rear wing of some tractors enable the driver to couple up mounted implements more easily by operating the hydraulic linkage while standing at the side of the tractor.

The instruction book should be studied to learn the correct operation of your tractor hydraulic system.

The Instrument Panel

There is a tremendous variation in the number and range of switches, gauges and coloured indicator lights on different makes, models and ages of farm tractor.

Plate 1.13   Analogue dials in the quiet cab on this tractor are used to register engine speed, fuel level and engine temperature.

Hour meter This records the engine running hours and indicates engine speed in rpm. Some hour meters, or proof meters, measure the actual hours run, but on most tractors it is usual for the meter to record running hours at about two-thirds of maximum engine speed. With the engine at full throttle, more hours will be recorded than are actually worked by the tractor.

Tractor forward and reverse speeds in each gear may also be shown on the meter, but a digital read-out on the instrument panel, which indicates the speed in each gear at a certain engine speed, is more common.

Gauges Older and more basic tractors have conventional dial gauges on the instrument panel. The number and purpose of gauges varies with the model of tractor. Engine oil pressure, engine coolant temperature, an ammeter to measure the rate of battery charge and discharge and a fuel gauge are the most common.

Indicator lights As well as indicator lights to warn the driver of low engine oil pressure and battery discharge, others are used to warn of air flow restriction from the air cleaner, low fuel level and low hydraulic oil pressure. Coloured warning lights also show when the parking brake is on, when traffic indicators are in use and when reversing lights, rear working lights, main headlamp beam, etc are on.

Electronic instrument panel Most large modern tractors display information electronically, either on an instrument panel or a small screen on the armrest of the driving seat. Digital displays, coloured indicator lights or bar graphs monitor tractor performance including engine speed, engine oil pressure and temperature, fuel and oil levels, parking brake engagement, main headlight beam, rear working lights and traffic indicators. There may also be switches for the road lights, traffic indicators, four-wheel drive engagement, gear selection, starter, air conditioning, radio controls, etc.

Plate 1.14 The electronic controls in this tractor cab include instrument panel, joystick control for the hydraulics, power take-off and forward/reverse shuttle, monitor/GPS screen, hydraulic spool valve control, lights and ignition, fuse box and passenger seat. (Massey Ferguson)

In-cab computers, standard on many tractors, monitor engine operation and provide technical information at the touch of a button. Typical digital read-outs provide engine, power take-off and forward speeds, wheelslip information, fuel used per hour and per acre or hectare, fuel reserve, total engine hours and when the next service is due. Details of in-cab satellite navigation, headland management and guidance systems are explained in Chapter 7.

Electronic controls can also manage repetitive headland or end-of-row operations by automatically disengaging the diff-lock, four-wheel drive and the power take-off when the hydraulic linkage is raised and re-engaging them when the implement is lowered back into work. Automatic four-wheel drive systems are programmed to disengage the drive to the front wheels when the tractor travels above a certain speed on the road to reduce tyre wear and fuel consumption.

Starting the Engine

Farm safety regulations require the driver to always be on the tractor seat before starting the engine. Many tractors have a safety start mechanism requiring either the gear levers to be in neutral or the clutch pedal to be pushed fully down before the starter motor will operate.

To start a diesel-engined tractor:

1. Make sure that the gear and shuttle levers on tractors with either a mechanical or powershift transmission are in neutral. Ensure that the hand brake is applied and the power take-off is disengaged.
2. Ensure that, if a stop control button is fitted, it is fully pushed in to the 'run' position.
3. On tractors with a mechanical transmission, depress the clutch to reduce the load on the starter motor and then use the starter button or turn the key to engage the starter motor. Many older tractors have a heater or excess fuel button to help start the diesel engine on cold days. Failure to use this aid can result in a discharged battery.
4. If the engine does not start within 30 seconds, release the key or starter button switch and allow the starter motor to come to rest and then try again.
5. If the tractor is fitted with a turbocharger, the engine should be idled for a time after starting and before stopping the engine to allow the turbocharger to slow down, particularly with older tractor engines.

To stop a diesel engine Depending on the age of the tractor, the engine is stopped with the stop control button or the ignition switch. This operates a solenoid which cuts off the fuel supply to the engine.

USING TRACTORS ON THE ROAD

The Highway Code must be obeyed when driving any tractor or agricultural vehicle on the public highway. The maximum legal UK road speed for an agricultural tractor is 40 kph (25 mph). Tractors equipped with a full suspension system and ABS brakes may travel at up to 60 kph (40 mph) depending on national traffic regulations. The maximum legal road speed for other trailed machines, for example a baler or a crop sprayer, is 32 kph (20 mph).

The following points must be observed when driving a tractor on the public highway:

- Make sure the brake pedals are locked together and the brakes are in good working order.
- The windscreen must be clean; make sure the screen wipers and washer are in good working order.
- Suitable mirrors must be fitted to the tractor to give the driver a clear view of the traffic behind, even when pulling a trailer.
- The steering and tyres must be in good condition.
- Road lights must be in working order at all times.
- Agricultural trailed equipment over a certain gross weight must be fitted with brakes and the brakes on trailed equipment in excess of this weight must also be connected to the tractor braking system. Agricultural trailers hauled at speeds in excess of 25 mph on the public highway must meet the full braking requirements for ordinary goods-carrying road trailers.
- Amber flashing road beacons must be used when driving on dual carriageway roads. Flashing beacons should be used on other roads to warn traffic of slow moving vehicles.
- Special regulations apply when taking wide or projecting loads on the highway. If the tractor lights and reflectors are obscured, there must be additional lights on the load. Marker plates are required for long mounted or semi-mounted implements and for wide or overhanging loads.

Road Traffic Regulations are very complex: the information given above is merely a guide. To ensure they are within the law, drivers of tractors and self-propelled machinery must find out the exact requirements for the equipment being used.

Other points to remember when using a tractor or self-propelled machine on the highway include:

- The driver must hold the relevant licence to drive the vehicle on the highway. Wheeled tractors and most other agricultural vehicles can be driven on a standard UK car licence, but rubber-tracked crawlers require a separate test and licence.
- The tractor must have valid road tax or an exemption certificate. There must be a valid insurance certificate and the vehicle registration plate must be clearly visible from the rear.
- When towing a laden trailer, make sure the load is secure.
- Mud or other materials must not be allowed to fall on the road. When this cannot be avoided, the law requires that it must be removed to avoid creating a hazard to other road users.

SAFE DRIVING AND IMPLEMENT HITCHING

In addition to the Road Traffic Regulations, there are requirements in the Health and Safety at Work Act and the Prevention of Accidents to Children in Agriculture Regulations which must be observed. Department for Transport official leaflets explain the requirements of these regulations; the following notes are only a guide to their content.

- Children under 13 may not drive or ride on a tractor or self-propelled machine or other farm machines.
- Young persons over 13 may only drive a tractor when under supervision and capable of driving safely.
- Young persons under 16 may only ride on a tractor or machine if a seat is fitted on the machine. Persons under 16 may not operate

complex power-driven machines such as a root harvester or a mechanical handler/loader.

- Passengers must not ride on the drawbar or other linkage on a tractor or any other machine.
- A tractor or other self-propelled machine may only be started or set in motion when the driver is seated in the normal driving position.
- The driver must not leave the seat while the tractor is in motion except in an emergency.
- Drawbar pins must be secured with a safety clip to prevent them jumping out while towing.
- All cutter bars must be covered with a strong rigid guard when not in use and when taken on the road.

Children are killed in farm accidents every year. Always take extra care when children are playing nearby. When you need to back a trailer, check there is no one behind before reversing.

Quiet cabs isolate the driver from the sound of children playing, and the noise made by other workers and their equipment. Take special care when driving in and around farm buildings and be prepared for an unexpected hazard around the corner.

Safe Hitching and Parking

- Back squarely up to a mounted implement when preparing to attach the linkage arms. Do not try to lift or drag an implement into position.
- Never stand between the tractor and implement when helping to hitch a mounted implement. Always work from the side. Remember the correct sequence for attaching the three-point linkage is left-right-top.

- Always make sure the hand brake is on when parking a tractor. Set the gear lever in neutral. Power shift transmissions should be in neutral or park. Always remove the ignition key when parking a tractor.
- Always lower loaders and mounted implements to the ground when parking a tractor.

Safe Driving in the Field

- Take special care when driving on sloping ground. Select a low gear before driving downhill and only change gear when the tractor is stationary.
- Make sure that nobody is in the way when moving off, lowering an implement or engaging the drive to a machine.
- Do not drive close to the edge of ditches or dykes. Stay clear of the edge of silage clamps when consolidating the contents with a tractor. Clamps that are full or almost full to the top should be fitted with side rails.
- Drive at a safe speed at all times. Good drivers do not need to apply the brakes fiercely as they are in full control of the tractor.

All agricultural workers have a legal responsibly to make full use of guards and other protective equipment on tractors and farm machinery. Guards must be in position when tractors and farm machines are being used. Broken or damaged guards must be reported to the employer so that repair or replacement can be arranged. The Health and Safety at Work Act states that a worker has a duty to ensure that while at work his or her actions do not affect their health and safety or that of other workers.

There are numerous publications on farm safety available from organisations such as the Health & Safety Executive. You should read them carefully and follow their recommendations.

CHAPTER 2

Tractor Engines

With few exceptions, farm tractors have a four-stroke, water-cooled diesel engine. Some older tractors have an air-cooled engine. Both types of engine are called internal combustion engines because fuel is burnt inside the cylinders to produce the necessary heat for them to work. In contrast, a steam engine is an external combustion engine and the required heat is created outside the cylinder by burning coal or wood to produce steam which is directed into the cylinder.

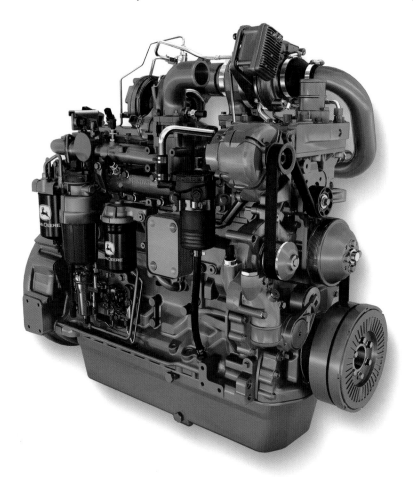

Plate 2.1   Selective catalytic reduction (SCR) technology to meet engine emission regulations, a common rail fuel injection system and a turbocharger are features of this 200 hp-plus diesel engine. (John Deere)

Single- and twin-cylinder four-stroke petrol engines, described in Chapter 8, are used on some farms to drive pumps, generators and other stationary equipment. Chain saws have small two-stroke petrol engines.

Diesel engines are more robust than petrol engines as they have to withstand much higher working pressures and temperatures. They offer greater efficiency and economy compared with a similar size petrol engine and develop full power very quickly when started from cold. Diesel engines are also known as compression ignition (CI) engines and petrol engines are sometimes called spark ignition engines.

THE FOUR-STROKE ENGINE

Four-stroke engines have one working or 'power' stroke in every four strokes of each piston. Every other stroke is a power stroke in a two-stroke engine.

In a four-stroke diesel engine, finely atomised fuel is injected into high temperature compressed air above the piston within an airtight cylinder. The piston is connected to a crankshaft.

The heat and resulting expansion caused by the burning fuel inside the cylinder drives the piston downward and turns the crankshaft.

Most tractors have a three-, four- or six-cylinder engine with an inlet and exhaust valve for each cylinder. However, some engines have two inlet and two exhaust valves in each cylinder. Air enters the cylinders of a diesel engine through the inlet valves. The exhaust valves release the waste gases to the exhaust pipe. The rocker arms, operated by pushrods and the camshaft, open the valves. The camshaft is driven by and timed with the crankshaft so that it opens the valves at the correct point in the four-stroke cycle. Strong springs close the valves. The camshaft gear, which has twice as many teeth as the crankshaft gear, completes one revolution for every two made by the crankshaft.

Multi-cylinder tractor engines have overhead valves which are above the pistons in the cylinder head and are opened when they are pushed downwards by the rocker arms. Some overhead engines have all of the valves at one side of the cylinder head. Others have a cross flow cylinder head with the inlet and exhaust valves on

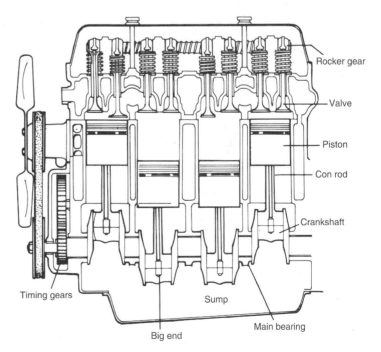

Figure 2.1   Section of a four-stroke overhead valve diesel engine.

opposite sides of the engine. This design has the advantage of a quick, straight-through exit for the exhaust gases and rapid entry for the incoming air through the inlet valves on the cooler side of the engine. Cold air is denser than warm air and this means that more oxygen will be drawn through the inlet valves into the cylinders.

Some single-cylinder engines have side valves (see Figure 8.1 in Chapter 8). Side valve engines, which are less efficient than overhead valve engines, have the valves in the cylinder block at one side of the pistons. The cam followers running on the camshaft open the valves by pushing them upward against the spring pressure and are closed again by the valve springs.

The piston is attached to the connecting rod (Figure 2.2) with a gudgeon pin, which runs in the little end bearing. The con rod is held on the crankshaft by the big end bearing. Each piston has a set of piston rings. The top rings are compression rings. They act as a seal between the piston and the cylinder wall to prevent loss of pressure above the piston on the compression stroke. The bottom ring is an oil ring. Some engines have two oil rings. Their purpose is to scrape oil from the cylinder wall to prevent it getting above the piston into the combustion chamber.

The crankshaft is secured to the cylinder block by the main bearings. A four-cylinder diesel engine has five main bearings. A heavy flywheel

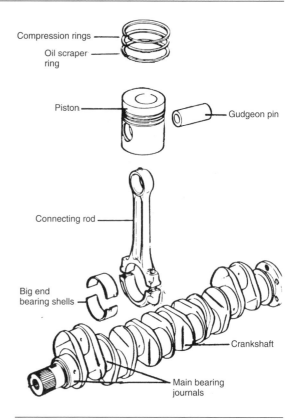

Figure 2.2   Piston and crankshaft components.

is bolted to the gearbox end of the crankshaft. It stores energy and keeps the crankshaft running smoothly between the power strokes. A timing gear, at the front of the crankshaft, drives the camshaft, and on a diesel engine it also drives the fuel injection pump. The camshaft is timed with the crankshaft to open the valves at the correct point in the four-stroke cycle. The injection pump is timed to inject fuel at the correct time on the compression stroke. Some timing systems are chain driven from the crankshaft.

All internal combustion engines have a cooling system to remove unwanted engine heat and almost all tractors have a water-cooling system. A few old tractors still in use have an air-cooled engine. Lubrication is vital as it limits the amount of heat and wear in an engine. All engines have a pressure lubrication system to ensure a constant flow of oil to all moving

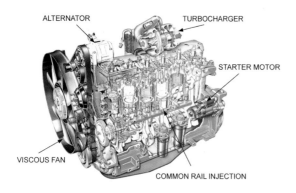

Plate 2.2   This six-cylinder diesel engine with a common rail fuel injection system has a cooling system with a viscous fan. (John Deere)

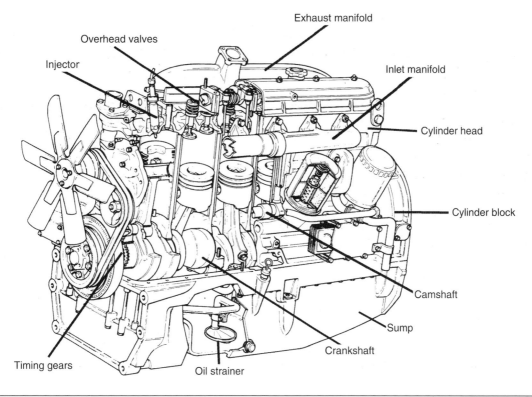

Figure 2.3   Sectional view of a diesel engine. (New Holland)

parts. The air cleaner is another essential part of an engine, providing a constant and copious supply of dust-free air which is mixed with the fuel in the cylinders; and the fuel system provides very precise amounts of fuel at the correct time. The following pages explain and illustrate the various components of a four-stroke diesel engine.

The Four-Stroke Diesel Engine

A diesel engine relies on the immense heat created inside the cylinder to ignite the fuel as it is injected into the cylinders. The high temperature is achieved by compressing the air in the cylinders to a very high pressure of 34 bar (500 psi) or more. This gives an air temperature in the region of 550° C in the cylinders. Because the heat created through compression ignites the fuel, a diesel engine is also known as a compression ignition engine.

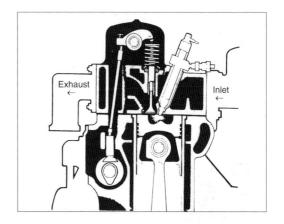

Figure 2.4   A direct injection cylinder head with a cross flow valve arrangement.

The four-stroke diesel engine works in this way:

Induction stroke The piston travels down and air is sucked from the air cleaner into the cylinder through the open inlet valve.

Compression stroke Both valves are closed. The rising piston compresses the air in the cylinder to a very high temperature. Just before the piston reaches the top of the stroke, a fine spray of diesel fuel is injected into the cylinder above the piston. The fuel burns instantly and the heat causes a rapid expansion of the gases above the piston.

Power stroke The expanding gases drive the piston down, turning the crankshaft. This is the working stroke of the four-stroke cycle.

Exhaust stroke The exhaust valve opens at the end of the power stroke. The waste gases are forced out of the cylinder by the rising piston. The inlet valve opens, just before the top of the exhaust stroke, in readiness for the next induction stroke.

A typical tractor engine runs at speeds well above 2,000 rpm, and at this speed each piston will make 4,000 strokes every minute. Each valve will open and close 1,000 times and each cylinder will receive 1,000 separate injections of diesel fuel.

Firing Order

Most tractor engines have three, four or six cylinders in-line, but there are exceptions including vee-engines with two rows of cylinders set at an angle to each other in the cylinder block. The firing order is the sequence in which the cylinders receive an injection of fuel. Number one cylinder is normally at the front of the engine. The firing orders for three-, four- and six-cylinder engines are:

- 3 cylinder 1.2.3
- 4 cylinder 1.3.4.2. or 1.2.4.3
- 6 cylinder 1.5.3.6.2.4 or 1.4.2.6.3.5

Compression Ratio

This is the relationship between the volume in the cylinder with the piston at top dead centre (TDC) and the volume bottom dead centre (BDC). The piston is at TDC when it is at the highest point in the cylinder and at BDC is when the piston is at its lowest point in the cylinder.

A typical compression ratio for a diesel engine is 16:1. This means that the volume in the cylinder at BDC is 16 times greater than the volume at TDC. This degree of compression creates the very high temperature required to ignite the fuel when it is injected in a very fine mist into the cylinder. A petrol engine, which employs a spark to ignite the fuel, has a compression ratio of about 8:1.

THE DIESEL FUEL SYSTEM

Direct and Indirect Injection

These are the two types of diesel engine used in agriculture. The direct injection engine has the fuel injected directly into the cylinder above the piston. This is the hottest part of the engine and gives the direct injection fuel system the advantage of a good starting performance, even on cold mornings.

The indirect injection engine has the fuel injected into a pre-combustion chamber at one side of the main combustion space above the piston. This design improves the efficiency of the engine because the pre-combustion chamber ensures that the fuel is thoroughly mixed with the air during the injection period. However, indirect injection engines can be difficult to start in cold weather and a heater or an excess fuel device or a combination of the two is used to help start a cold engine.

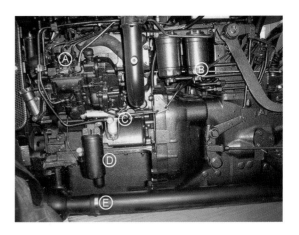

Plate 2.3   An example of a diesel engine fuel injection pump (A), fuel filters (B), fuel sediment bowl (C), engine oil filter (D), drive shaft to front wheels (E).

The Fuel System

The function of the diesel fuel system is to inject the correct quantity of clean, atomised fuel into the cylinders at the correct time in the four-stroke cycle. The components of a fuel system vary with different types of diesel engine. The amount of fuel injected will depend on the speed of the engine. Typically the diesel fuel flows through a tap under the tank to the fuel filter. The tap is always left on to prevent bubbles of air forming in the fuel. Air in the fuel can cause misfiring and in any quantity will stop the engine. Air will enter the fuel system if the engine runs out of fuel and when the tap is turned off to service the fuel system. The engine is unlikely to restart until the air has been bled (removed) from the fuel (see page 19).

The lift pump Some engines have a lift pump, usually driven by the camshaft, which delivers fuel under slight pressure through the fuel filters to the injection pump. The lift pump may have a filter to trap any water and dirt present in the fuel and a hand-priming lever to bleed air from the fuel system. Fuel systems without a lift pump rely on gravity feed to the filters and the injection pump.

The fuel filter removes all traces of dirt from the fuel and some have a water trap to collect any water that may occur through condensation in the tractor fuel tank. Some engines have two fuel filters. Fuel filters must be renewed periodically usually after 600 hours or at the service indicated in the operator's manual. A bleed screw on the filter unit is used to remove any air trapped in the fuel system after renewing the filter element or if the tractor runs out of fuel.

Fuel Injection Systems

Many farm tractors have a rotary (DPA) fuel injection pump, which supplies minute quantities of diesel fuel to each fuel injector at very high pressure and at the correct point in the four-stroke cycle. Some older diesel-engined tractors have an in-line fuel injection pump. To meet the requirements of current engine emission regulations, many of the newer tractors have a more efficient common rail injection system with a single pumping element and electronically operated injectors. This has the advantage of better fuel economy, lower engine noise and faster engine response to changes in load.

Injection pump Driven by the engine timing gears, or timing chain, the injection pump delivers diesel fuel to each injector at a pressure of approximately 175 bar (1 bar = 14.7 psi). Modern high-performance engines may have injection pressures approaching 1400 bar or more. There are two types of injection pump: rotary (or DPA) and in-line.

A *rotary or DPA pump* has a single pumping element with a distributor head, which supplies fuel to each injector in the engine firing order. This design ensures that all cylinders receive equal quantities of fuel even when the pumping element is worn.

An *in-line pump* has a separate pumping element for each injector. The pump is set by the manufacturer to ensure that each injector receives an equal quantity of fuel. A worn pump will not do this efficiently, which results in wasted fuel.

The throttle setting regulates how much fuel is injected into the cylinders, and in this way controls engine speed. A governor is built into the injection pump. This maintains the engine speed at the level selected with the hand or foot throttle control. The governor allows the injection pump to supply extra fuel when the engine speed falls away as a result of an increased load on the engine. When the load is reduced and the engine speed is too high, the governor reduces the amount of fuel supplied by the injection pump to the injectors, bringing the engine back to the required speed.

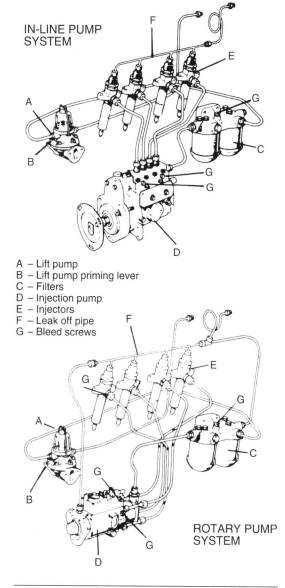

IN-LINE PUMP
SYSTEM

A – Lift pump
B – Lift pump priming lever
C – Filters
D – Injection pump
E – Injectors
F – Leak off pipe
G – Bleed screws

ROTARY PUMP
SYSTEM

Figure 2.5   In-line and DPA fuel injection systems.

is normally used for indirect injection engines with pre-combustion or swirl chambers that help to mix the injected fuel with the high temperature air in the cylinder. The injection pump supplies more fuel than is needed by the injectors. The excess fuel is used to lubricate and cool the injectors and is then returned to the tank by the injector leak-off pipes.

Failure to supply clean fuel to the injectors will cause an increase in the rate of wear. Worn injectors will inject too much fuel into the cylinders. This is a waste of fuel. A sure sign of worn injectors is excessive quantities of black smoke coming from the exhaust pipe.

Common rail injection system In-line and DPA injection pumps pressurise the fuel and control the timing of each injection into the cylinders. Common rail (Plate 2.3) differs in that a pressure pump, protected from damage by a relief valve, provides a supply of high-pressure fuel to the common rail or pipe which serves as the fuel reservoir. Electronically operated injectors inject minute quantities of fuel from the common rail into the cylinders and a computer controls injection timing, fuel pressure and the amount of fuel injected. Depending on engine speed the common rail injection system operates at injection pressures of between 200 and 1500 bar or more.

Care of Fuel System

When a tractor is kept outside for long periods, there can be a problem with water condensing inside the fuel tank and contaminating the fuel. Refuelling at the end of the day will reduce condensation inside the tank. Always use clean fuel. If there is any doubt about its cleanliness when refuelling from a can, use a funnel with a fine mesh filter.

Regular servicing of fuel filters is very important because, with time, the filter element will be partially blocked with dirt and will not work efficiently. For most diesel engine fuel systems, the filter element should be renewed after every 600 hours of service or as recommended in the operator's instruction book. Always use a new

Injectors The injector, or atomiser, sprays a fine mist of fuel thorough a minute hole or holes in the tip of the injector nozzle into the cylinder, when the piston is just a few degrees before TDC on the compression stroke. Direct injection engines usually have multi-hole injector nozzles to help mix the fuel with the pressurised air in the cylinders. A single hole nozzle

rubber sealing ring when renewing a filter element. Many fuel filter elements have a built-in sealing ring. Always check the operator's manual to find the filter change period for your tractor. Some filter units have a water trap at the bottom of the filter element, and this should be drained periodically. The injection pump or the common rail injection system is unlikely to require any attention between major overhauls.

A trained mechanic should check the injectors with an injector tester at intervals of about 1,000 hours, but this period will vary with different engines. Excessive black smoke from the exhaust pipe is a sign of faulty or worn injectors; this wastes fuel and they should be replaced with re-conditioned service exchange units.

Some in-line pumps have an oil level plug which should be checked periodically.

Bleeding the Fuel System

This may be necessary after renewing the filters or if the engine runs out of fuel, as there will be air in the fuel system. The engine may not start without first bleeding the fuel system to remove this air. Always wear rubber gloves or use a suitable barrier cream when working with diesel fuel. This will provide protection from dermatitis, an unpleasant skin inflammation that can result from handling tractor fuel.

With plenty of fuel in the tank and the tap turned on, slacken or remove the bleed screw in the top of the filter unit. Some engines have a priming lever on the fuel lift pump which is used to pump fuel from the slackened bleed screw until it runs free from air bubbles. Tighten the bleed screw, and if the engine has two fuel filters, repeat the process. Engines without a priming lever are bled by gravity – slacken the bleed screw and let the fuel flow until it is free from air bubbles. If there is a bleed screw on the injection pump, repeat the process until the fuel is free from air bubbles.

Engines with a rotary pump may not start without first bleeding the injector pipes. Slacken or remove at least two of the pipes from the injectors. Run the engine with the starter motor until fuel is seen coming from the pipes. Replace and tighten the injector pipe unions. The engine should now start and will soon run smoothly. (See Figure 2.5 for location of the bleed screws.)

Some common rail injection systems have a hand-priming lever to bleed the fuel system after servicing. On other engines the starter motor is used to remove any air trapped in the fuel system.

Cold Starting

Diesel engines can be difficult to start in cold weather and for this reason many engines have some form of cold starting aid. Use of a cold starting aid is usually necessary for indirect engines and may be required on some direct injection engines in severe weather conditions. The main types of cold starting aid are:

Excess fuel device Extra fuel can be supplied to an engine with an in-line injection pump by pushing in the excess fuel button before starting a cold engine. The button returns to the normal running position when the engine has started. Some injection pumps have an excess fuel device which automatically supplies extra fuel when starting a cold engine. Rotary fuel injection pumps may have an automatic mechanism that retards the timing and the fuel is injected at a slightly later stage on the compression stroke.

Heaters A small heater coil in the inlet manifold warms the air before it enters the cylinders. The heater is usually operated with the tractor ignition switch or key. Another type of heater plug is fitted in each of the pre-combustion chambers of some indirect injection engines. The heaters warm the air automatically when the engine is started, and on some engines they remain hot for a few seconds after the engine has started. Some older diesel engines have a cold starting aid consisting of a heater element combined with an excess fuel unit in the inlet manifold.

Storage of Diesel Fuel

Diesel fuel is clean when it is delivered to the farm and should be stored in a way to keep it clean and ensure the fuel injection system has a long and trouble-free working life.

Approved tanks, usually constructed from medium-density polyethylene or non-galvanised steel, are suitable for storing diesel fuel. The tank, often on a stand to facilitate refuelling by gravity, should be sited for easy access for both tractors and farm machines and for delivery tankers.

To prevent pollution, the tank must either be double skinned or surrounded by a bund wall large enough to hold the contents of a full tank plus 10 per cent allowance for rain water. Liquid drained from the bund must not be allowed to pollute a watercourse.

Plate 2.4 This free-standing fuel storage tank, with an electric pump to refuel tractors, has a double skin to prevent any leakage of fuel that might result from any accidental damage.

A ground level tank with a top outlet and an electric pump for refuelling tractors reduces the risk of pollution. A small farm may have a 2,500 litre capacity diesel tank while large farming estates may have a tank holding 40,000 litres or more.

A sludge cock at the lower end of the tank is used to drain off water and sludge deposits before the tank receives a new delivery of fuel. The draw off pipe for filling a tractor tank must have a tap that automatically closes when it is not in use. When closed, the filler opening at the top of the tank should be airtight with a vent pipe allowing air to enter the tank while restricting the entry of dust and water. When possible, a new delivery of fuel should be left to settle before any fuel is drawn off.

Biodiesel

With minor modifications, some diesel engines will run on up to 100 per cent biodiesel made from oilseed rape. Bioethanol made from wheat or sugar beet can, in some cases, be used as a substitute for diesel fuel. Some tractors have separate tanks for diesel and biofuel; they can run on either fuel or a mixture of both fuels depending on local availability.

Red mineral-based diesel has been the standard fuel for diesel-engined farm tractors for many years, but to help meet the current exhaust emission regulations, diesel fuel now has a much lower sulphur content. The fuel may also have a small biodiesel content derived from rape seed oil.

Exhaust Gas Emissions

EU regulations limit the level of harmful exhaust gas emissions from a diesel engine which contain harmful particulates – minute particles of unburnt fuel which form soot. The regulations also limit the level of nitrous oxide gas (NoX), which adds to the level of greenhouse gases in the atmosphere. When introduced in 1996, the regulations limited the level of particulates and

NoX which could be emitted by tractors over 175 horsepower.

Further restrictions have come into force since that date, reducing by 97 per cent the level of particulates and NoX that new tractor diesel engines made since 2014 can emit compared with 1996 levels. Engine modifications to meet these regulations include the use of computerised engine management systems, more efficient engine cooling and variable geometry turbochargers. The recirculation and cooling of exhaust gases before they are released to the atmosphere also achieves a reduction in the level of harmful engine emissions.

Tractor manufacturers use either Exhaust Gas Recirculation (EGR) or Selective Catalytic Reduction (SCR) engine technology to help meet emission regulations, the latter in conjunction with Diesel Exhaust Fluid (DEF), often known by the trade name AdBlue.

The EGR system extracts and cools a portion of the gas before it leaves the exhaust manifold and mixes it with incoming clean air being drawn into the engine. This reduces the volume of fresh oxygen entering the cylinders, which in turn lowers the combustion temperature and reduces the quantity of harmful nitrous oxide in the exhaust emissions. These engines also have a variable geometry turbocharger, which provides the required quantity of air to ensure full combustion and complete burning of the fuel.

The SCR process involves the injection of small quantities of DEF, a solution of aqueous urea, into the catalyst chamber that forms part of the exhaust system. The fluid, which is stored in a separate tank, is sprayed into the exhaust gas leaving the cylinders, where it converts the NoX in the exhaust gases into nitrogen and water before they are released from the exhaust pipe. The engine management system automatically prevents the engine from running without a supply of DEF.

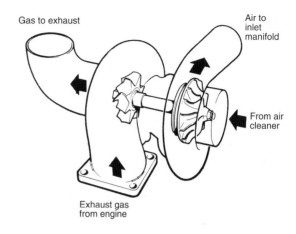

Figure 2.6   Turbocharger.

TURBOCHARGERS

The power developed by an engine is determined by the amount of fuel it burns during the brief combustion period. There must be sufficient air to enable the engine to burn all of the fuel as completely as possible, and a turbocharger helps to maximise the amount of air taken into the cylinders.

The turbocharger, mounted on the engine, is a form of air blower. The hot exhaust gases drive the turbine (fan) as they pass from the engine to the exhaust pipe. A second fan, on the same shaft, draws air from the air cleaner. This fan – or turbine – forces the incoming air under pressure into the cylinders and by increasing the volume of air, the engine will burn more fuel. The turbocharger shaft runs at about 20,000 rpm with the engine at idling speed and up to 80,000 rpm at full throttle. When turbocharged, the power of a 75 kW (100 hp) engine is increased to about 100 kW (130 hp). It is important to service the air cleaner at frequent intervals on a turbocharged engine.

Variable Geometry Turbocharger

A variable geometry turbocharger is often used on high horsepower tractor engines. It has variable angle vanes on the turbine which regulate the air pressure in the cylinders. The engine management system closes down the vanes to boost air pressure at low engine speeds and opens them at high speeds to limit the maximum air pressure in the cylinders.

Wastegate Turbocharger

Turbochargers deliver vast quantities of air to the cylinders in order to increase the power output at high engine speeds, but they tend to supply too much air when the engine runs near its maximum speed.

A wastegate turbocharger provides increased power at the lower end of the speed range of an engine. It also overcomes the problem of supplying too much air at high engine speed by reducing the speed of the turbocharger fan. This speed reduction is achieved by diverting some of the exhaust gases through a wastegate valve before they reach the turbine fan.

In Figure 2.7 the turbocharger (B) supplies pressurised air to the cylinders from the air cleaner (A). Exhaust gases rotate the turbine as they pass through it on their way to the exhaust pipe (E). When the engine runs at full speed and the air pressure in the manifold reaches a pre-set level the wastegate valve (D) is opened by diaphragm (F). This allows some of the gases to bypass the turbine and limit the speed of the turbocharger. When the manifold pressure decreases, the spring (C) closes the wastegate.

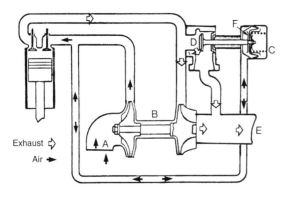

Exhaust
Air

Figure 2.7   A wastegate turbocharger.

Intercooler

Cool air is denser than warm air. So, by cooling the pressurised air after it leaves the turbocharger, even more air can be forced into the cylinders to increase engine power.

An intercooler, or aftercooler, usually found on very high horsepower tractors, is used to reduce the temperature of the air before it reaches the cylinders. There are two types. An air-to-air intercooler uses air from the engine-cooling fan to reduce the temperature of the air after it leaves the turbocharger. An air-to-water intercooler cools the air by passing it through a radiator at the front of the engine.

Power Boost

Some electronic tractor engine management systems provide power boost above certain engine speeds. This automatically increases the fuel supply to bring the engine back to speed and power when the tractor is subjected to an increased load, either at the power take-off or when doing heavy haulage work causes the engine speed to fall away. Power boost also reduces fuel supply when the engine speed exceeds the pre-set level. The percentage of power increase varies with different tractors and gear selection. Engine power boost is typically between 9 and 17.5 per cent above the basic horsepower rating.

AIR CLEANERS

All modern tractors have a dry element air cleaner, but some older models still in use have an oil-bath air cleaner.

The Pre-Cleaner

Pre-cleaners remove large pieces of dust and dirt before the air reaches the main air cleaner element. There are various types of pre-cleaner. They include a fine mesh or gauze screen over the air cleaner inlet to trap large particles of dirt, centrifugal pre-cleaners and exhaust aspirated systems.

Centrifugal pre-cleaners have a dome-shaped cover with angled inlet louvres which cause the air to circulate inside the dome. Centrifugal force causes the heavier particles to be thrown outwards and they are either collected in a transparent bowl attached to the air inlet or discharged from a slot in the upper surface of the dome.

Exhaust aspirated pre-cleaners have an inlet duct which causes the incoming air to spin and throw the heavier particles outwards. They are drawn from the air cleaner housing by a vacuum tube connected to the exhaust system and discharged with the exhaust gases.

The engine air supply is drawn through a thick paper or felt filter element on its way to the cylinders. Most tractors have a gauge on the instrument panel to warn the driver when the airflow to the engine is restricted.

A dry air cleaner can be cleaned by tapping the element gently against a clean tractor tyre. This must be done gently to avoid denting the element, which will be useless if damaged. The element should be rotated between each tap on the tyre.

As an alternative, compressed air may be used, but the air pressure should not exceed 2 bar (30 psi). The air should be used to blow from the inside of the element outwards. Do not hold the air nozzle too close to the element as this may damage it.

The inner filter element should not be removed. If the air cleaner restriction warning

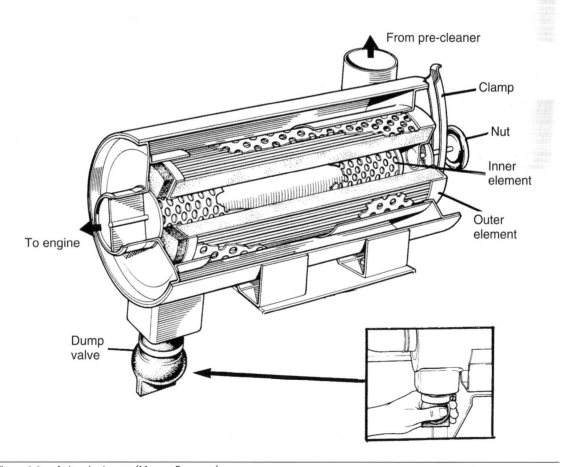

Figure 2.8   A dry air cleaner. (Massey Ferguson)

light still glows after servicing the outer element the inner section should be replaced.

A dry air cleaner filter element should be checked and cleaned weekly or at the service intervals given in the instruction book. In dusty conditions the element should be checked every day. Air cleaner maintenance is very important. A clogged air cleaner will starve the engine of air, resulting in power loss, fuel wastage and undue wear.

Oil-Bath Air Cleaner
Some older tractors have an oil-bath air cleaner. Dry air cleaners are suitable for most conditions, but an oil-bath cleaner is more efficient where dust is a serious problem.

Most oil-bath air cleaners have a dome-shaped pre-cleaner to remove large particles of dirt. The air then passes through an oil-impregnated fine metal gauze on its way to the engine cylinders. When the engine stops, the oil drips back into the oil bath and any dirt collected in the oil is deposited in the oil bath.

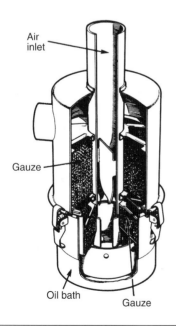

Figure 2.9   An oil-bath air cleaner.

The oil level in the oil bath should be checked every 50 hours (or weekly), but more frequently when working in dusty conditions. When there is about 12 mm of dirt in the bottom of the oil bath it should be cleaned out and re-filled to the level mark with clean engine oil.

ENGINE LUBRICATION

Metal surfaces cannot be machined to a perfectly smooth state, so when two metal surfaces run together at speed, there will be a great deal of friction, resulting in rapid wear and excessive heat. By introducing a lubricant film between the two metal surfaces, there will be much less wear and the oil will disperse most of the heat.

It is vital to remove much of the heat produced to ensure the moving parts do not overheat or, in extreme situations, seize. Both the lubrication and cooling systems play an important part in removing unwanted engine heat.

The three main functions of engine oil are:

1. To lubricate the moving parts and reduce friction.
2. To help cool the engine by removing some of the heat caused by friction.
3. To form a seal between the pistons and cylinder walls. Without this seal there would be very little compression in the combustion space.

Oil Classification
Traditionally, oil has been classified according to its viscosity or thickness. Different oils are given a viscosity number; thin oils have a low number and thick oils have a higher one. The viscosity numbers used for lubricating oil were devised by the American Society of Automotive Engineers (SAE). The viscosity range includes SAE 20 and SAE 30 for an engine oil and SAE 40 and SAE 50 for transmission oil.

The viscosity of even the best quality lubricating oils will change with variations in temperature. However, modern oils have additives that keep the viscosity stable with changes in the temperature of the oil.

Multi-grade engine oils, which span a range of viscosities, are used for diesel engines. They have a dual SAE rating. For example, a 10W-40 engine oil for diesel engines has an SAE 40 rating when the engine is hot and a rating of SAE 10 when it is cold. This oil is thin enough for starting on a cold morning yet thick enough to be efficient at normal operating temperatures. These oils must also withstand the very high running temperatures of turbocharged engines.

Additives

Viscosity rating refers to the thickness of an oil, and depending on their use, most agricultural oils contain a variety of special additives to ensure a long life for the oil and for the engine. These additives include:

Detergent stops a harmful build-up of deposits which are produced when fuel is burnt and cause a build-up of sludge in the oilways and bearings. When the oil is drained, the sludge deposits are carried away with the oil.

Dispersants keep the soot, formed when fuel is burnt, in suspension in the oil, avoiding sludge deposits in the oilways, piston ring grooves, etc.

Anti-oxidant reduces the effects of oxidation (heat and chemical breakdown), which thickens the oil.

Anti-corrosive protects the engine from the corrosive action of water and acids formed when fuel is burnt.

Anti-wear reduces the wear rate of moving metal surfaces, especially when the engine is under heavy load.

Viscosity index improver thickens engine oil at high operating temperatures.

Anti-foam This prevents the formation of foam in the oil which occurs when air is absorbed by the lubricant. It is a particularly important additive for oil used in hydraulic systems.

Oil Contamination

The efficiency of lubricants, especially engine oils, is reduced with use. Diesel engines require an oil change at frequent intervals, which depending on the model of tractor may be between 250 and 500 hours of service. Always consult the operator's manual to find the correct oil change period for a particular tractor engine. The oil must be drained and renewed because with use it will become contaminated with:

Sludge is formed from carbon, water and other substances when fuel is burnt in an engine. It will block oilways and bearings and, in time, will clog piston rings.

Dust enters the engine and adds to the build-up of sludge, especially when the air cleaner and crankcase breather are not serviced regularly.

Metal particles, including minute pieces of metal from the bearings and other parts of the engine, will find their way into the oil, especially when the engine is new.

Engine heat causes the gradual breakdown of the structure of the oil. Chemical changes also occur due to the effects of oxygen (oxidation).

Water, produced when fuel is burnt, and also from condensation, adds to oil contamination. A yellow sludge on the underside of the oil filler cap is an indication that the oil may be contaminated with water.

Dirty oil filters, poor storage for new oils and dirty oil handling containers will also add to the problems of oil contamination. These can be avoided with good storage, regular servicing and clean utensils.

Changing Engine Oil

The oil should be drained when the engine is hot. Most tractor engines have a spin-on disposable canister-type filter element which should be removed at the same time. Elderly tractors may have an oil filter bowl with a replaceable element. The old element should be scrapped and the filter bowl thoroughly cleaned before fitting a replacement.

After draining the old oil, replace the sump plug, using a new washer if the existing one is damaged. Refill the sump with the correct grade of new engine oil to the full mark on the dipstick. Run the engine for a short while then re-check the oil level on the dipstick. Finally, check for leaks and record the engine hourmeter reading so you will know when the next oil change is due.

Engine Lubrication Systems

Four-stroke engines are lubricated by a force feed system with a pump that circulates oil to the moving parts. A splash feed system is used for some small single-cylinder engines. Two-stroke petrol engines used for chain saws and other

applications have an oil mist lubrication system using a mixture of petrol and oil in the fuel tank.

Pressure lubrication A force feed or pressure lubrication system has a pump driven by the engine crankshaft, camshaft or timing gears. Gear and rotor pumps are the most common designs. A gear pump has two gears meshed together inside a housing. One gear, attached to the drive shaft, turns the second gear; and oil, which enters one side of the pump housing, is carried between the gear teeth to an outlet in the opposite side of the housing.

A rotor pump works in a similar way but with a moving rotor turning inside a fixed rotor ring. The moving rotor has one less lobe than the fixed rotor ring and oil is carried in the spaces between the lobes on the moving rotor

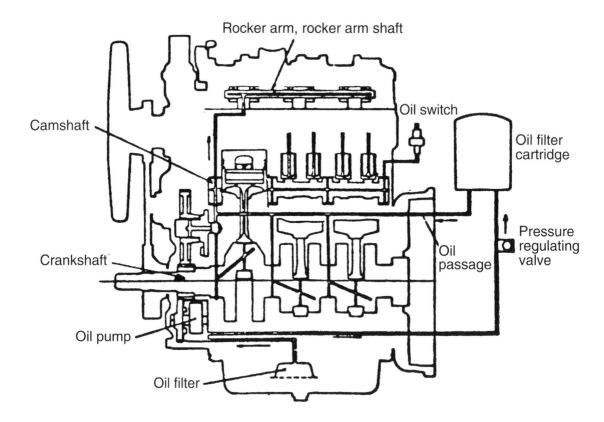

Figure 2.10   A four-stroke engine lubrication system.

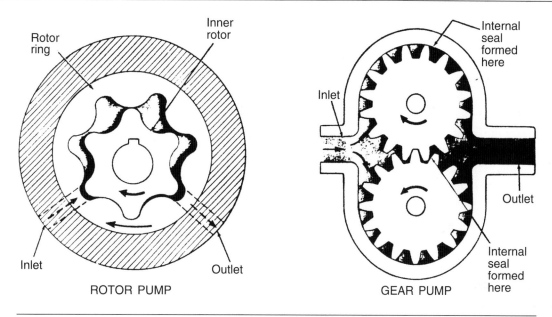

Figure 2.11   Types of pump used to circulate engine oil.

to the outlet point, where it is forced from the pump housing.

The pump draws oil from the sump through a mesh strainer and pumps it to a filter element, which removes the smallest particles of dirt from the lubricant. The filtered oil is pumped along oil galleries, or drillings, in the crankcase and moving parts to lubricate the main bearings, big end bearings and the valve gear. Oilways, or drillings, in the connecting rods carry oil to the little end bearings. The oil dispersed inside the engine by the crankshaft and other moving parts lubricates the cylinder walls and other components. Some engines have a pressure spray system to lubricate and cool the bottom end of the pistons (skirts) and the lower part of the cylinder walls. The valve pushrod ends and other valve gear components are lubricated by the oil as it drips back down into the sump for recirculation.

A relief valve, usually a ball and spring arrangement, is included in the system to protect the pump from overloading, particularly at high engine speeds when the oil pressure may exceed the pre-set limit. When the relief valve blows, the oil bypasses the system and is returned direct to the sump.

There are two types of pressure lubrication system. The full flow system filters the total flow of oil from the pump before it is circulated to the bearings. The bypass lubrication system, found in older tractor engines, has two oilways from the pump outlet. Some of the oil passes through the filter and returns to the sump, while the remainder is pumped direct to the moving parts. The bypass system will still lubricate the engine if the filter is blocked through lack of maintenance.

Oil cooler Some of the more powerful tractors have an engine oil cooler. This is usually in the form of a separate small radiator attached to the cooling system radiator. Some engines have a heat exchanger unit attached to the engine block and connected to the engine-cooling system.

Crankcase breather This ventilates the sump to prevent a build-up of pressure and the formation of water-laden gases in the sump. This can occur in a worn engine where gases escape past the piston rings and down into the sump. The breather, sometimes in the oil filler cap, has a small filter element to prevent dust entering the sump. This should be cleaned when changing the engine oil.

Oil warning light Connected to a pressure switch in the main oil gallery, this is included in all pressure lubrication systems. The light should go out when the engine is started. Stop the engine immediately if the light remains on or appears at any time when the engine is running. Locate and rectify the fault before re-starting the engine.

Oil pressure gauge Some tractors have a gauge on the instrument panel which either measures the actual oil pressure or indicates that the pressure is within safe working limits. A typical oil pressure is 2.8 bar (40 psi), but this will drop when the engine is hot. Very high or low oil pressures should be investigated.

Causes of low oil pressure include:

1. Low oil level.
2. The engine oil is too thin. This may be because:
   (a) The engine is overheating.
   (b) The oil has been diluted with fuel.
   (c) The wrong grade of oil is being used.
3. Worn oil pump or worn engine bearings.
4. The relief valve is stuck open or has a weak spring.
5. Blocked oil filter (but only if the pressure gauge is after the filter in the oil flow circuit).

Causes of high oil pressure include:

1. The relief valve is stuck in the closed position.
2. The oil is too thick. This may be because:
   (a) The engine is running too cold.
   (b) The oil is very dirty.
   (c) Use of the wrong grade of oil.
3. Blocked oil filter only when the gauge comes before the filter in the oil circuit.

COOLING SYSTEMS

Transfer of Heat

Heat can be transferred for one point to another in three different ways:

Conduction Hold a piece of metal in a fire and before long the end held in the hand becomes warm. Heat moving through metal in this way is transferred by conduction. Metals, especially copper and brass, are very efficient conductors of heat.

Convection Hot air will rise above cold air because it is less dense and therefore lighter. Hot water, like air, also rises. A domestic convector heater works in this way.

Radiation Rays of heat can be felt as they leave a fire. These rays always travel in straight lines from the heat source. This form of heat transfer is called radiation.

When an engine is cooled, the heat is conducted through the cylinder walls into the cooling water. Convection occurs as the hot water rises up through the cylinder block to the radiator, where the water is partly cooled by radiation. Heat is also radiated from the exhaust manifold.

Engines create a vast amount of heat when fuel is burnt and this must be removed to ensure efficient engine performance. Most tractors are water-cooled but a few of the older tractors still in use have an air-cooled engine.

Plate 2.5   A tractor cooling system with separate radiators for air conditioning (A), intercooler (B), transmission oil (C) and the engine water-cooling system (D).

Water Cooling

Water cooling partly depends on the thermo-syphon principle whereby hot water, which is lighter than cold water, rises by convection to the upper part of the engine and is then cooled as it passes down through the radiator. A belt-driven pump assists water circulation in the cooling system.

The layout of a basic tractor engine water-cooling system is shown in Figure 2.12. The water takes up engine heat as it rises up through the water jacket in the cylinder block; and the fan, which draws cold air through the radiator, cools the water as it flows down through the radiator.

The radiator has a series of small water tubes surrounded by air passages and fins. Most radiators have a pressurised cap, which, by increasing the working pressure in the cooling system, raises the boiling point of the water. A pressure valve in the radiator cap prevents the escape of the cooling water at normal working pressure but allows water and steam to escape when the pressure exceeds its pre-set level. A typical tractor engine-cooling system will have a cap with a valve which opens at a pressure of 5.5 bar (7 psi).

Some tractors have a small expansion tank connected by a rubber tube to the radiator. On rare occasions when it is necessary to top up the cooling system, water is added to the expansion tank.

Pressurised radiators without an expansion tank have an overflow pipe in the radiator neck. When the temperature and pressure in the cooling system are higher than normal, the pressure valve in the radiator cap will open to release the pressure, sometimes in the form of water and steam from the overflow pipe.

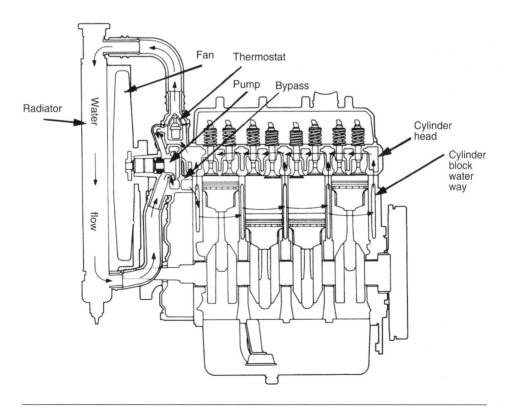

Figure 2.12   An engine water-cooling system.

To avoid scalding, care must be taken when removing the radiator cap from an engine. If it is necessary to remove the cap before the engine has cooled down, it should be removed slowly, preferably with a piece of thick material placed over the radiator cap.

The fan, usually attached to the water pump, is driven by a vee-belt, or fan belt, driven from the crankshaft pulley on some tractors. Others have a variable speed viscous drive fan with a thermostatic coupling between the drive pulley and the fan blades. The viscous coupling matches the fan speed to the engine's cooling needs by running at high speed when the engine is hot and at a reduced speed when the engine runs at lower operating temperatures.

A variable speed fan gives a faster warm up from cold, helps to reduce exhaust emissions to meet engine operating regulations and absorbs less engine power when the tractor is used for light and medium duties. Some cooling systems have a reversible fan which is used to clean dirt from the radiator by blowing it from the front of the radiator core.

The water pump, or impeller, assists water circulation round the cylinder block and cylinder head. On most tractor engines it is driven by the fan belt, but on some engines the engine timing gears drive the water pump.

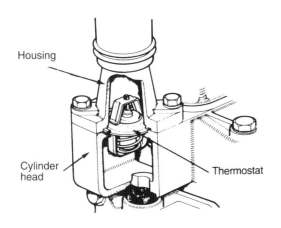

Figure 2.13   The thermostat in an engine-cooling system.

The thermostat provides automatic control of engine temperature. Situated in the hottest point of the cooling system in the housing at the front of the cylinder head, it is connected by a flexible rubber hose to the top of the radiator. When the engine is cold the thermostat is closed to prevent water circulating through the radiator and to bring the engine to its operating temperature as quickly as possible. When the thermostat is closed, the cooling water, which removes heat as it rises up through the engine to the water pump, is returned through the bypass hose (Figure 2.12) back to the bottom of the cylinder block. When the water reaches its operating temperature of about 75° C, the thermostat valve opens so that the hot water can flow down through the radiator to be cooled.

Traditionally, tractor engines have one of the three types of thermostat – bellows, bi-metallic and wax or pellet.

Bellows thermostats contain a liquid such as alcohol or ether with a low boiling point. When the liquid boils, it expands the thin metallic bellows connected to the thermostat valve. This then opens, allowing the hot water to circulate through the radiator to be cooled.

Bi-metallic strip thermostats have coils made with steel and bronze strips that are bonded together and connected to the thermostat valve. Bronze expands at a faster rate than steel when heated, so as the water temperature increases the bronze expands at a faster rate than the steel, causing the bi-metallic strip to uncoil and open the thermostat. When the water cools, the coil contracts and closes the valve.

Wax or pellet thermostats have a solid wax pellet in a small sealed container attached to a plunger, which in turn is connected to the thermostat valve. As the water in the cooling system reaches its operating temperature, the wax melts and it expands the plunger and the thermostat valve opens. When the water cools, the wax re-solidifies and closes the thermostat valve.

Most cooling systems have two drain taps, one in the cylinder block and the second at the bottom of the radiator.

Engine Operating Temperature

A cold or an overheated engine cannot work efficiently. The cooling water should be at a temperature of approximately 90° C for efficient operation. The temperature gauge on the instrument panel should be monitored at all times.

Frost Protection

An anti-freeze solution, ethylene glycol, is used to protect the cooling system from frost damage. A 25 per cent mixture of anti-freeze and water – 2 litres of anti-freeze in 8 litres of water – will protect the system down to a temperature of about 18° C.

Anti-freeze does not prevent freezing; it only lowers the freezing point of the coolant. It also lowers the boiling point. When necessary, it is best to top up the cooling system with a ready-mixed anti-freeze solution during the winter months. Anti-freeze contains an additive or inhibitor to minimise corrosion in the cooling system.

Care of the Cooling System

- Check the coolant level in the expansion tank, and when required, top it up with the correct mixture of water and anti-freeze.
- The radiator water level in cooling systems without an expansion tank should be checked from time to time, especially in very hot weather.
- The water should be about 25 mm below the radiator neck to allow for expansion when the engine is hot.
- Never add large quantities of cold water to a hot engine when it is very short of coolant. Cold water may crack the cylinder block.
- Keep the outside of the radiator fins free from dirt and chaff. Remember to blow or wash the dirt out the way it came in by using an airline or hosepipe working from the fan side of the radiator.
- Maintain the correct fan belt tension as instructed in the operator's manual.

Causes of Engine Overheating

- Low water level.
- Radiator blocked. This can be either inside or outside the radiator.
- Slack fan belt.
- The thermostat is stuck shut.

Air Cooling

Most small engines and a few older tractors are air cooled. There is no risk of frost damage with an air-cooled engine, but it may overheat if the cooling fins are not kept clean.

Air-cooled tractor engines have a separate blower that creates a continuous blast of cold air over the cooling fins around the cylinders, which helps to disperse engine heat. However, air-cooled engines are no longer used for farm tractors, as they fail to meet current EU exhaust emission regulations.

Small single-cylinder engines have fan blades on the flywheel, which disperse engine heat by blowing air over the cooling fins on the cylinder head and block.

CHAPTER 3

Tractor Electrical Systems

The tractor battery provides the electricity required to operate the many items of electrical equipment found on tractors and self-propelled machines. It also supplies the power for the remote control systems on various implements and machines. An explanation of the principles of electricity is given in Chapter 21.

THE BATTERY

A battery stores direct current (DC). Mains electricity is alternating current (AC) and cannot be stored in a battery unless it is first converted (rectified) to DC current. A mains-operated battery charger rectifies AC current to DC current and at the same time transforms it from mains voltage to the standard 12 volt current used for tractor electrical systems.

Tractor batteries, which are of the lead acid type, store electricity generated by the tractor alternator. Some older tractors may have a dynamo that serves the same purpose.

A battery consists of a polypropylene box containing a number of lead plates immersed in sulphuric acid. There are two terminal posts on a battery. Current flows from the alternator through the positive terminal in to the battery, and a cable from the negative terminal to the tractor chassis provides the earth connection. Electricity from the alternator is converted into chemical energy by the battery and is stored in that form. During the charging process, oxygen and hydrogen are given off and the strength of the battery electrolyte (sulphuric acid) increases. The acid strength of the electrolyte will decrease as the battery discharges.

Most tractor batteries are sealed and, apart from keeping them clean, require little maintenance. Some have a cap with a ventilation hole for each cell in the battery. The caps release oxygen (O) and hydrogen (H) gases given off when the battery is being charged. This removes water (H_2O) from the electrolyte which must be replaced, when necessary, by topping up the battery cells with distilled water.

When the starter or other electrical equipment is used, the stored chemical energy is converted back to electricity. The charging system provides most, if not all, of the electrical current needed to operate the electronic control systems, lights, horn, screen wipers, radio, etc when the engine is running at normal operating speed. The battery provides backup when the engine idles and the considerable amount of power required by the starter motor when starting the engine.

Battery Maintenance

The following maintenance points will help to ensure long trouble-free service from sealed polypropylene-cased batteries:

- Make sure that the terminal connections are clean and tight. A light coating of petroleum jelly will reduce corrosion.
- The earth strap from the negative terminal to the tractor chassis must be tight and corrosion free.
- Keep the top of the battery clean and dry. It is possible for a battery to discharge itself over a period if the top is damp and dirty.
- The battery must be secure in its mounting. This will prevent damage to the casing.

- A discharged battery will freeze. It is a sensible precaution to remove the batteries from equipment when it is stored in an open building and not used for long periods during the winter.
- The negative earth cable should be disconnected first when removing a battery. It should be re-connected last when installing it.

Additional maintenance for batteries with removable cell caps:

- Keep the vent holes in the caps clear to allow gases to escape when the battery is charging.
- Remove the cell caps at regular intervals to check the electrolyte level, which should be just above the plates. If it is low, use distilled water to restore the electrolyte to its correct level. Tap water should not be used as it contains impurities and may shorten battery life.
- Remember to wipe any dirt from the top of the battery before removing the caps and wipe away any distilled water spilled while topping up the cells.

The state of charge of a battery with removable caps can be checked with a hydrometer. It measures the specific gravity (concentration)

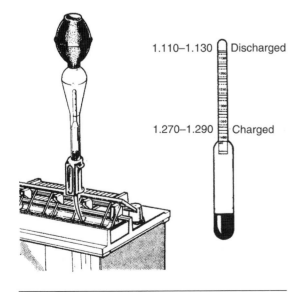

Figure 3.1   Testing a battery with a hydrometer.

of the sulphuric acid. A float in the hydrometer body has either three coloured bands or a scale to indicate the strength of the electrolyte. The coloured bands indicate full charge, half charge or discharged.

The specific gravity scale readings are:

Specific gravity	State of charge
1.280	Full charge
1.200	Half charge
1.110	Discharged

To use a hydrometer, remove the cell cap and put the rubber tube into the electrolyte. Squeeze the rubber bulb and then gently release it to draw some electrolyte into the hydrometer glass or transparent plastic body. Note the state of charge, or specific gravity reading, and return the liquid to the cell. Repeat the procedure for at least two more cells. Remember that the electrolyte is acid, so keep it away from your clothes and skin. After use, rinse the hydrometer in clean water.

Battery Charging

A discharged battery can be recharged with a mains battery charger. This converts mains AC current to DC current, which can be stored in a tractor battery. The simplest type of charger, commonly known as a trickle charger, is connected to a 13 amp socket. The charging rate is very low, typically 2 or 3 amps, and it will take between 12 and 24 hours to recharge a battery. Large farm workshops may have a heavy-duty charger that will recharge a battery very quickly. Both types of charger have an adjustment which offers the choice of a 6 or 12 volt output. Many trickle chargers have a 2 volt setting, and some heavy-duty chargers have a 24 volt option.

Follow this procedure to charge a battery with removable cell caps:

- Make sure the top of the battery is clean. Remove the cell caps and top up the electrolyte if necessary. Leave the cell caps slack; this will prevent a build-up of gas during the charging process.

- Connect the charger clips to the battery terminals, making sure that the negative and positive clips are on the correct terminals. Check that the charger is set to suit the battery voltage.
- Plug the charger into the mains supply and switch on.
- When a trickle charger is used, check the electrolyte every few hours with a hydrometer. When fully charged, switch off the charger at the mains supply before removing the terminal clips. Tighten the cell caps.
- Because a battery gives off hydrogen and oxygen when under charge, avoid the use of naked flames in the area. *Never used a lighted match to check the cells.*

Using Jump Leads

Tractor batteries are sometimes discharged when the starter switch is used excessively. This problem can be overcome with a pair of jump leads, but if the battery is faulty, it must be replaced. Jump leads are very heavy cables with a connecting clip at each end.

Be careful when using jump leads, as failure to follow the correct procedure may result in personal injury or damage to the vehicle's electrical system. Follow this sequence:

1. Connect the positive terminals on both batteries with one jump lead.
2. Connect the other jump lead to the negative terminal on the booster battery.
3. Using a good earth point on the tractor, which should be at least 500 mm from the battery, connect the opposite end of the negative lead.
4. Make sure that the leads are clear of any moving parts and then operate the starter.
5. When the engine has started, remove the jump leads in the reverse order.

It is a dangerous practice to connect the jump lead to the negative terminal on a discharged battery. This action can result in severe arcing, which could have explosive results. Never allow the jump lead clips to touch together when the opposite ends are connected to the booster battery.

CHARGING SYSTEMS

An alternator or a dynamo is used to generate the electricity which is stored in a chemical form in the tractor battery. An alternator is standard equipment on all modern tractors, but some older models have a dynamo. The main advantages of an alternator are its high output at low engine speed and a greater output compared with a dynamo of the equivalent size.

Electricity is generated when a magnet rotates inside a number of coils or windings which form a static ring of electrical conductors. An alternator has static conductor windings (stator) and a rotating magnet (rotor) which generates alternating current (AC). This is then rectified (converted) to direct current (DC) before it is stored in the battery.

In a similar way, electricity is generated when an electrical conductor is rotated in a static electro-magnetic field. This principle is used in a dynamo that has fixed magnets and a rotating conductor (armature) which generates DC current.

The Alternator

Driven by the fan belt from the engine crankshaft pulley, an alternator has fixed conductor

Plate 3.1   Two vee-belts are used to drive the alternator from the crankshaft pulley on this tractor.

windings (stators) and rotating magnets (rotors). Current is supplied from the battery to the rotor windings by two carbon brushes on the rotor shaft to create an electro-magnetic field. Carbon is a very good material for conducting electricity. Alternating current, induced (created) in the stator windings by the rotating magnets, flows to a rectifier in the alternator housing, which converts it to direct current before it is stored in the battery. When an engine runs at 2,500 rpm, the alternator will be driven at speeds of 6,000 rpm or more.

The Dynamo

A dynamo has an armature (rotor) with a large number of insulated electrical conductors which rotate at speed in a magnetic field created

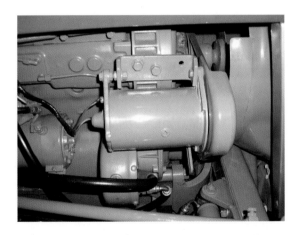

Plate 3.2   Some older farm tractors have a dynamo for battery charging.

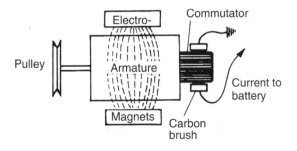

Figure 3.2   The principle of generating electricity with a dynamo.

by two diametrically opposite electro-magnets attached to the dynamo housing. There is some residual magnetism in the magnets, but current supplied by the battery, known as the field current, creates and controls the strength of the magnetic field. The electricity generated when the armature rotates in the magnetic field passes from the commutator at one end of the armature shaft to two carbon brushes held against the commutator by spring pressure. One of the carbon brushes collects the current from the commutator and feeds it to the battery. The second carbon brush is connected to earth.

The Control Box

It is necessary to limit output from the generator when the battery is fully charged and prevent reverse flow from the battery when the engine is running at a slow speed or stopped. The control box is part of the charging circuit and contains the voltage regulator and cut out or relay switch, which carry out these functions.

The voltage regulator in the alternator charging circuit allows current to flow to the stator windings but restricts the flow when the battery is fully charged. Alternators also have an automatic relay switch. Its purpose is to prevent back flow from the battery when the engine is running at low speed or is stopped.

The voltage regulator in a dynamo charging circuit prevents overcharging by regulating the strength of the field current to the magnets. It provides a strong field current to give a high charging rate when the battery is only partly charged and reduces the field current when it is fully charged. The cut out is an automatic switch which prevents reverse flow from the battery to the dynamo when the engine runs at a very slow speed or is stopped.

Simple Charging Faults

The ignition warning light on the instrument panel warns the driver that the alternator or dynamo is not charging. It may also come on when the engine is running at a very low speed, but it should go out as soon as the engine revs are increased. If the ignition warning light comes on when the engine is running normally, there is no need to stop the engine immediately except when the fan belt has broken. However, the fault should be investigated as the battery will soon become discharged, especially if headlights and other electrical equipment are being used.

Some simple reasons why an alternator or dynamo does not charge include:

- Loose terminal connections or a broken wire.
- Broken fan belt or one with the tension too slack.
- Worn, dirty or broken carbon brushes.

Maintenance of the Charging System

Correct tensioning of the drive belt is very important. Most alternators and dynamos are driven by the fan belt from the engine crankshaft pulley. Some engines have a separate drive belt for the alternator. A slack fan belt can cause engine overheating as well as a low rate of charge. An over-tightened belt will, in time, damage the generator bearings.

To adjust the fan belt, slacken the securing bolts and move the generator on its pivots until the correct tension is achieved. Finally, re-tighten the bolts and check tension.

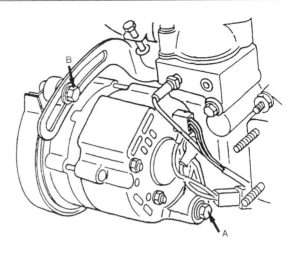

Figure 3.4 Fan belt adjustment. Slacken the alternator pivot bolt (A) and the adjuster stay bolt (B), then move the alternator outwards to increase belt tension and re-tighten the bolts. (Massey Ferguson)

Alternators and dynamos have an internal fan on the rotor or armature shaft to keep them cool by drawing in air through ventilation louvres. They must be kept clean to allow maximum airflow. No other attention is needed, apart from checking that the terminal connections are clean and tight.

STARTING SYSTEMS

The electric starting system has a powerful motor with a small gear, or pinion, which turns

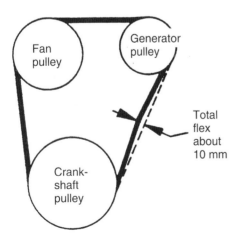

Figure 3.3 Checking fan belt tension. The total flex on the belt when correctly tensioned should be about 10 mm.

Plate 3.3 A diesel engine with a pre-engaged starter motor (A) and solenoid (B).

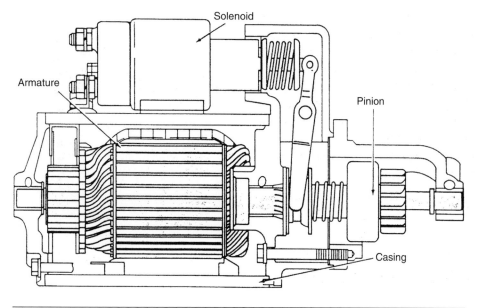

Figure 3.5   Sectional diagram of a pre-engaged starter motor.

the engine flywheel. All modern farm tractors have a pre-engaged starter, but old petrol/TVO engined tractors have an inertia starter.

Pre-engaged starter motors have a solenoid which engages the starter pinion with a large ring gear on the flywheel. The solenoid, operated with the starter key or switch, also connects the battery to the motor. A starter motor turns a diesel engine at approximately 200 rpm.

A solenoid consists of an electro-magnetic coil around a soft iron plunger in a sealed housing. There are two terminals on the end of the solenoid housing with heavy-duty cables; one is connected to the battery and the other to the starter motor. When the starter key or switch is used, the battery energises the solenoid coil to create a magnetic field which pulls the plunger against spring pressure through the centre of the coil towards the flywheel ring gear. After the solenoid plunger has meshed the starter pinion with the flywheel ring gear, it connects the two solenoid terminals to complete the circuit between the battery and the starter motor.

When the engine starts and the switch is released, the coil is no longer energised and the

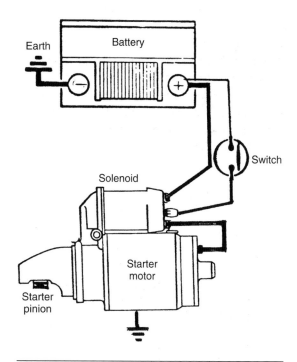

Figure 3.6   Circuit diagram for a pre-engaged starter motor.

solenoid spring returns the plunger to the rest position and in so doing cuts off the current and withdraws the starter pinion from the flywheel

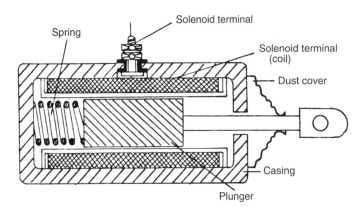

Figure 3.7 Sectional diagram of a solenoid. When the battery energises the solenoid, the plunger moves forward against the spring and meshes the starter pinion with the flywheel ring gear. The spring disengages the pinion when the starter switch is released.

ring gear. Starter motors require a high flow of current and should only be operated for a few seconds at a time.

The only driver maintenance required is to keep battery, solenoid and starter motor terminals clean and tight.

Another type of pre-engaged starter motor, found on some older tractors, has a hand lever used to engage the starter pinion with the flywheel ring gear before energising a solenoid to complete the electrical circuit to the starter motor.

Inertia starter motors are not suitable for diesel engines but are sometime used on ATVs and small compact tractors. When the starter switch is used, a small pinion moves along a helix (very coarse thread) on the motor shaft to engage with the ring gear and turn the engine. When the engine starts, the pinion is thrown out of mesh by the faster turning ring gear, the current is cut off and the pinion returns to rest.

LIGHTS AND FUSES

Tractors are equipped with a full set of road lights that comply with the Road Traffic Regulations. Many tractors have additional front and rear working lights for working in the field after dark. Direction indicators and hazard warning lights are also standard equipment. All lights should be checked frequently and faulty bulbs replaced. It is a legal requirement that the lights must be in working order at all times.

All tractors have fuses to protect the electrical circuits from overload. A fine wire in the

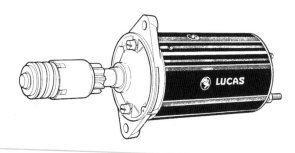

Figure 3.8 The pinion on an inertia starter motor moves along its shaft into mesh with the flywheel ring gear.

Plate 3.4 A pre-engaged starter motor. The solenoid moves the starter pinion into mesh with the flywheel ring gear and then connects the two terminals on the solenoid to complete the circuit from the battery to the starter motor.

fuse melts (blows) if too much current flows through the circuit. This prevents damage to various items of electrical equipment such as electronic transmission and hydraulic system controls, lights, heater, windscreen wipers, radio, etc.

The flow of electric current in a circuit is measured in amps. A fuse is rated according to how many amps the fuse wire will carry before it blows. Try to find out why a fuse has blown before replacing it and always use a fuse of the same rating as the original. A bigger fuse will not give the proper protection to the equipment and for this reason should not be used.

Plate 3.5   A tractor with full set of lights and orange flashing beacon. (John Deere)

CHAPTER 4

Tractor Transmission Systems

Tractor engines run at speeds in excess of 2,000 rpm. In the lowest gear, the rear wheels turn at 20 rpm or less. The transmission system reduces the engine crankshaft speed by means of the gearbox, crown wheel and pinion and final drive to provide a suitable forward speed for the work being done.

In a low gear, a tractor can pull a heavy load, but the engine would stall if the driver attempted to pull the same load in a high gear. The gearbox provides a range of forward speeds and pulls to suit all types of work from deep ploughing to high-speed haulage work and a range of reverse speeds.

Plate 4.1 A sectional view of a four-wheel drive tractor showing the line of drive from the engine to the front and rear wheels. (John Deere)

In simple terms, the load pulled by the tractor multiplied by the forward speed is equal to the power required for a specific task. When used in the field, the load on the drawbar or hydraulic linkage will determine the gear used and forward speed. A heavy load such as a plough will require a much lower gear than a light draft hay tedder or fertiliser broadcaster.

A tractor can pull a much heavier load at a higher speed on the highway than it can in the field because the hard road surface offers little resistance to forward travel compared with loose or wet soil.

The increased traction of a four-wheel drive tractor enables it to pull a greater load than a two-wheel drive model of the same horsepower, and almost all farm tractors in current production have four-wheel drive. This extra power can, for example, be used to pull wider implements, reducing the number of passes across a field and the overall cost of growing a crop. The front wheels on most four-wheel drive tractors are about 30 per cent smaller in diameter than the rear wheels, but the wheels on very high horsepower tractors tend to be of equal size to maximise traction, pulling power and weight distribution.

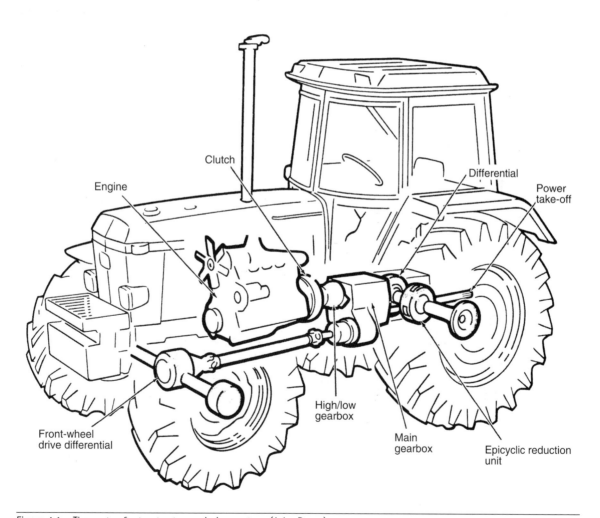

Figure 4.1   The parts of a tractor transmission system. (John Deere)

There are several types of tractor transmission system. Some older models have a simple manual gearbox with six or eight forward and two reverse gears. Farm tractors now have a much wider selection of gears which, depending on make and model, range from 12 to 40 forward speeds and in some cases the same number in reverse. Many have a semi- or fully automatic powershift transmission where operation is automated. Other tractors have a CVT (constantly variable transmission) system with a single clutchless speed band.

Line of Drive

The engine flywheel is connected to the gearbox by the clutch, which is used to engage the drive to a manually controlled transmission with gear levers to select forward and reverse gears. Drive from the gearbox is transferred through ninety degrees by the crown wheel and pinion, which also carries the differential and reduces the speed of the shafts driving the rear wheels. The differential allows the wheels to turn at different speeds when the tractor turns a corner. Power from the differential is transmitted by the half-shafts to the rear wheels. Many tractors have final reduction units, which give a further speed reduction before the power reaches the wheels.

The power take-off is driven from the gearbox, and many tractors have front and rear power take-off shafts. The gearbox also drives the hydraulic pump on some tractors while others have the hydraulic pump mounted on and driven by the engine. This makes the hydraulic system completely independent of the transmission.

CLUTCH

The clutch, which is used to engage and disengage drive from the engine to the transmission when changing gear on tractors with a manually controlled gearbox, also enables the driver to move off smoothly.

Many farm tractors with a single or dual dry plate clutch are still in use. Advanced transmissions with semi- or fully automatic,

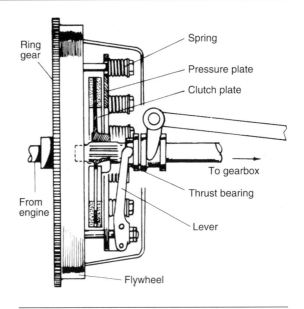

Figure 4.2   A single plate clutch assembly.

change-on-the move gearboxes used on many of the more powerful tractors have hydraulically operated, oil-immersed, multi-plate wet clutches. Many tractors also have independent multi-plate hydraulic clutches to engage and disengage drive to the power take-off and to the front wheels on four-wheel drive models.

Single Plate Clutch

Most single plate clutches on older tractors are dry, but some models have a wet clutch which is similar in construction but with a different type of friction lining designed to run in oil.

The main clutch assembly, with a smooth pressure plate and a set of springs, is bolted to the engine flywheel and, with the exception of the clutch disc, which is attached to a splined shaft from the gearbox, it rotates with the flywheel.

When the clutch pedal is up in its disengaged position and the tractor is in motion, the clutch disc is held against the flywheel by the pressure plate and clutch springs and drive is transmitted to the gearbox. When the pedal is pushed down, a linkage releases the spring pressure on the clutch disc and breaks the drive to the gearbox. The pressure plate and springs continue to turn with the flywheel but the clutch disc is stationary.

Almost without exception, tractors have a live power take-off. This means that the drive to the power take-off shaft is not disconnected when the clutch disengages the drive to the rear wheels.

There are two ways of achieving live power take-off drive:

Independent clutch Tractors with an independent clutch for the power take-off have a single-stage transmission clutch. The multi-plate power take-off clutch in the transmission housing is hydraulically operated with a push button or a lever in the cab.

Dual clutch Although found on some older tractors, the dual clutch has been replaced by an independent power take-off. A dual clutch works in the same way as a single-stage clutch, but there are two clutch discs, two pressure plates and two sets of springs. When the pedal is pushed halfway down, spring pressure is released from the transmission clutch disc and disconnects drive to the wheels. Pushing the pedal fully down to engage the second stage of the clutch stops the power take-off.

Strong coil springs are used to hold the friction disc against the flywheel, but some tractors have a clutch with a Belleville spring. This is similar to a very large concave washer made from spring steel. When the clutch pedal is depressed, the Belleville spring is squeezed flat and disengages the drive. When the clutch pedal is released, the spring returns to its concave shape, grips the friction plate against the flywheel and re-engages the drive.

Another type of clutch has a diaphragm spring to hold the friction disc against the flywheel. A hydraulic release mechanism disengages the drive by pushing the centre of the diaphragm spring towards the engine, which in turn withdraws the pressure from the pressure disc and breaks the drive to the gearbox.

Clutch Pedal Free Play

The clutch pedal must have free movement before any spring resistance is felt; this ensures the clutch is fully engaged.

Clutch pedal free play is maintained automatically on most tractors with a hydraulically operated clutch, but it may be necessary from time to time to adjust the pedal free play on a mechanically operated clutch. Check for free play by pushing the pedal down by hand. A typical free play measurement is 18 mm but this will vary with different models of tractor. Never use the clutch pedal as a foot rest because this will take up all of the free play and cause unnecessary wear on the clutch thrust bearing.

Too little free play can result in:

1. Wear on the clutch thrust bearing.
2. Clutch slip, because the pressure plate does not hold the clutch plate firmly against the flywheel.

Too much free play will make gear engagement difficult because the drive between the friction disc and the pressure plate is not fully disengaged.

Oil-cooled multi-plate hydraulic clutches are used on many tractors, and those with a change-on-the-move transmission have separate clutches to engage forward and reverse

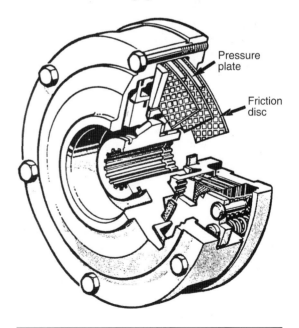

Figure 4.3   A multi-plate wet clutch with three friction plates. (John Deere)

ratios. This type of clutch has a pedal-operated master cylinder (a small pump and fluid reservoir) which supplies oil to a small hydraulic ram that actuates the clutch linkage. When the pedal is depressed, the ram releases the spring pressure on the clutch plate and disengages the drive. Drive is re-engaged when the pedal is released and the fluid returns to the reservoir.

Multi-Plate Wet Clutch

The life of a single plate clutch is limited by the area of the friction faces and its efficiency in dispersing the heat created by friction when the clutch is engaged or disengaged. Clutch life also depends on the type of work done by the tractor and the way it is driven. Rapid clutch wear will result from the constant forward and reverse action over many hours when using a front-end loader, a problem which can be overcome by using a tractor with a forward and reverse shuttle unit.

Hydraulically operated and oil-cooled multi-plate transmission clutches are much more efficient at dispersing heat and are used on many of the more powerful models of farm tractor. There are several designs of multi-plate clutch, but all have alternate driving and driven plates held together by hydraulic pressure provided by the tractor hydraulic system.

Inching Pedal

Some tractors with powershift transmissions have an inching pedal instead of a clutch pedal. This is a wet, multi-plate clutch pack actuated by an electro-hydraulic mechanism, which is controlled with a pedal or switch. It is used to over-ride the hydraulic drive mechanism when hitching an implement or driving in confined spaces.

Torque Converter

Some tractors have a fluid coupling (or torque converter) which consists of a pump, a turbine and a stator situated between the engine flywheel and the clutch.

The pump supplies oil at a high rate but at relatively low pressure to the turbine blades on a rotor attached to the clutch input shaft. After the oil has passed through the turbine, a set of fixed vanes on the stator change the direction of the oil flow and return it to the pump. After selecting the required gear, the tractor will move off when the driver uses the hand or foot throttle. The tractor will continue to accelerate with further throttle movement until it reaches the highest possible speed in the selected gear. The throttle controls oil flow in the turbine and the engine power is not transmitted to the gearbox until the oil flow reaches a pre-set level.

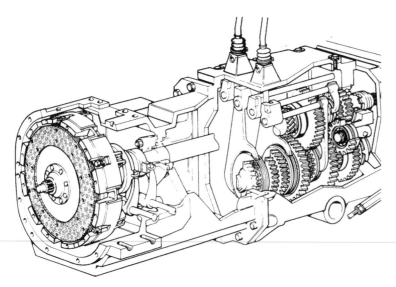

Figure 4.4   The output shaft from the dry plate clutch in this transmission system is connected to a sliding mesh gearbox with straight tooth spur gears. The selector forks, which move the gears, are also shown. (New Holland)

GEARBOXES

The gearbox provides a range of forward and reverse speeds and a neutral position. The gear selected not only determines the maximum forward speed but also how much the tractor can pull. Depending on the gear selected, forward speeds can range from 1 to 30 kph (0.6–20 mph), while more powerful tractors have a top speed of 40, 50 or even 60 kph. An optional creeper gearbox is sometimes used for specialist rowcrop work that requires a very slow forward speed.

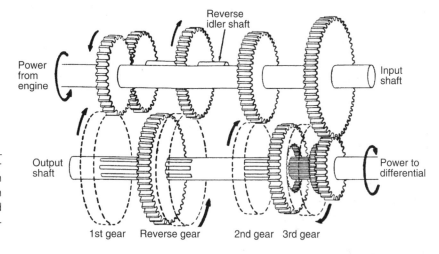

Figure 4.5   The gear arrangement in a simple sliding mesh gearbox with straight tooth gears. The gear on the third idler shaft is used when engaging reverse.

Plate 4.2   With a suitable braking system, tractors used for road haulage work have a top legal speed of 25 mph. (John Deere)

Various types of gearbox are used on farm tractors. The most basic mechanical gearbox has six or eight forward speeds and two reverse gears. More complex gearboxes with many different combinations of forward and reverse gears are in use to allow for slow field speeds or fast road transport work. Some have twelve gears in both directions; others have many more, ranging from 16 × 16 up to 60 × 60 in forward and reverse. Change-on-the-move gearboxes including power shuttle and synchro-shuttle for forward and reverse operation and CVT (Continuously Variable Transmission) are standard or optional equipment for many tractors.

A basic tractor gearbox has two levers; one selects the required forward or reverse gear and the second range lever engages high or low ratio. It is usually necessary to put one of the levers into neutral before the tractor can be started. On some, a safety mechanism to ensure the tractor cannot be started in gear means that the clutch pedal must be depressed before using the starter switch.

An additional switch or lever, controlled hydraulically or electrically, is a further refinement on some tractors. This allows the driver to select from two to eight ratios in each gear while on the move without using the clutch.

A constant mesh gearbox found on older tractors may have either spur gears or quieter running helical gears (see Figure 4.6), which have greater strength. Small groups of gears on the top and bottom shaft are always in mesh, but only transmit drive when they are linked together by splined hubs or couplers. The gear lever and selector forks move the couplers on splined shafts to connect the groups of gears in different sequences. Constant mesh gearboxes usually have a synchromesh unit between each group of gears. A synchromesh coupling has a pair of male and female cone-shaped hubs, which bring the couplers on the two groups of gears to the same speed.

Change-on-the-move gearboxes are used on many farm tractors. They range from a mechanical gearbox with a forward/reverse shuttle to a

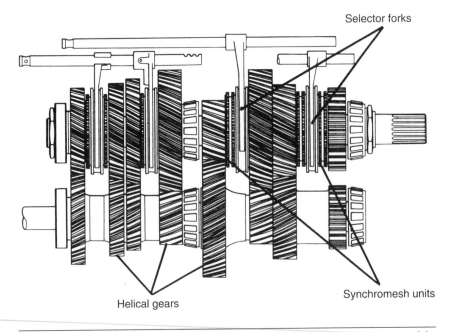

Figure 4.6 A synchromesh gearbox with helical gears. Different gears are engaged by connecting the various groups of gears. (New Holland)

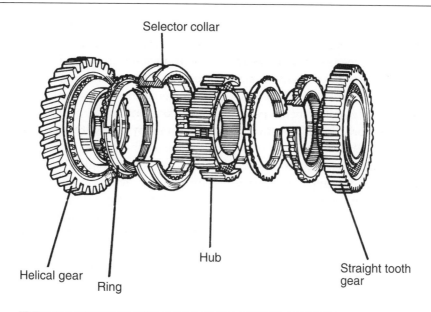

Selector collar

Hub

Helical gear

Ring

Straight tooth gear

Figure 4.7   Cone-shaped connecting collars in this synchromesh unit bring two gears shafts to the same speed before they are brought into mesh. (New Holland)

Plate 4.3   Front and rear power take-off shafts and hydraulic linkages, plus a wide range of forward and reverse gears, make this tractor equally suitable for both land work and high-speed road haulage.

fully automatic or CVT transmission with completely stepless changes in speed.

Many tractors have a semi- or full-powershift transmission, allowing gear changes to be made on the move. With a semi-powershift gearbox, the clutch is used to change between groups or ranges of speeds, and any of the speeds in each range is selected by simply moving a lever without the need to use the clutch. A full-powershift system progresses the tractor through the gear changes without any input from the driver.

Continuously variable transmission (CVT), used on some larger tractors, combines hydrostatic (oil) and mechanical (gear) drive systems for completely stepless travel with no need for the driver or an electronic system to change gear.

In all of these systems, changes between forward and reverse are usually achieved with a shuttle lever, which may or may not require

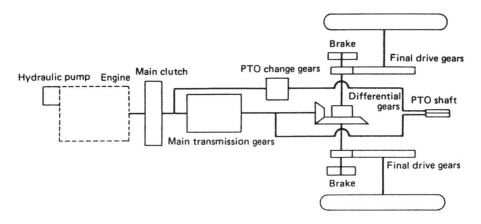

Two-wheel drive layout with the final drive gears located near the rear wheels.

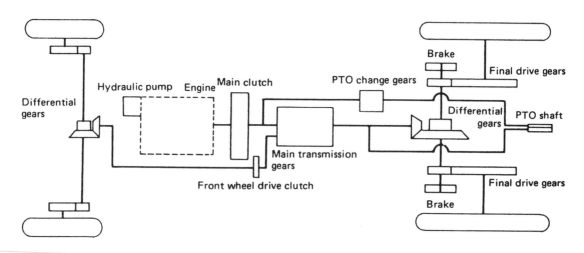

Layout of a four-wheel drive tractor transmission.

Figure 4.8   Examples of typical two-wheel and four-wheel drive transmission systems.

the use of the clutch, and engages a matched reverse speed.

REAR AXLE

The Crown Wheel and Pinion

The drive from the gearbox must be put through a right angle because the wheels run at 90 degrees to the engine crankshaft. The crown wheel and pinion bevel gears have the stronger and quieter bevel helical teeth, but some older tractors may have straight tooth bevel gears. The crown wheel and pinion change the direction of drive. The small bevel pinion (Figure 4.9) on the output shaft from the gearbox drives the much larger crown wheel. When a small gear drives a larger, there is always a reduction in speed. This means that not only does the crown wheel and pinion turn the drive through 90 degrees, but it also gives a considerable reduction on speed.

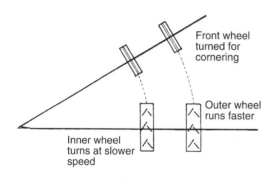

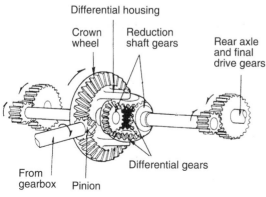

Figure 4.9   How a differential works. (Kubota)

DIFFERENTIAL

If the axle, or axles on a four-wheel drive tractor, were solid shafts with the wheels secured to them, it would be impossible to turn a corner with a tractor. The differential allows one wheel on the rear axle to turn faster than the other does when turning a corner, but when on a straight course the differential causes both wheels to run at the same speed.

Figure 4.9 shows a differential unit attached to the crown wheel. A bevel gear on each half-shaft, which transmits drive to one wheel, meshes with the smaller differential pinions running in housings attached to the crown wheel.

When the tractor travels in a straight line, the crown wheel, half-shaft and differential gears all turn as a single assembly. When a tractor turns a corner, the inner wheel slows down. As a result, the differential gears move round the slower half-shaft gear. This action increases the speed of the half-shaft gear connected to the outer wheel. Because this half-shaft gear turns faster, the outer wheel also turns faster as the tractor turns the corner. As soon as the tractor is back on a straight course, the half-shaft and differential gears stop turning at different speeds and both half-shafts run at the same speed.

When one wheel spins at high-speed and the other is stationary, the differential allows one half-shaft to run at twice the normal speed and the other does not move at all.

The Differential Lock

Known as the diff-lock, this device is used to prevent differential action by locking the two half-shafts together making it impossible for one wheel to spin when working in difficult conditions or if the tractor is overloaded. The diff-lock is engaged with a foot pedal or an electro-hydraulic switch and disengaged with a switch, lever or a brake pedal.

FINAL DRIVE

Many tractors have speed reduction gears between the crown wheel and the rear wheels.

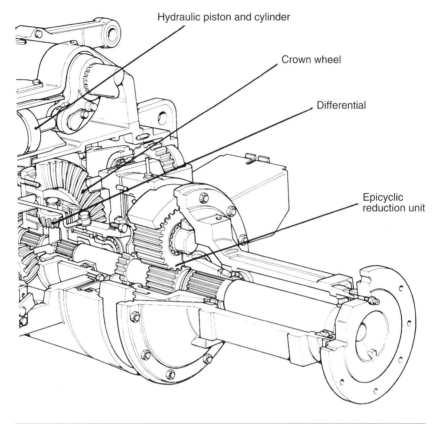

Hydraulic piston and cylinder

Crown wheel

Differential

Epicyclic
reduction unit

Figure 4.10   Layout of a transmission system with an epicyclic final drive unit. (New Holland)

Some tractors have two pairs of reduction gears – a small one driving a large one – inside the transmission housing, but most tractors have planetary or epicyclic reduction units built into the driving axles. A planetary unit (Figure 4.11) consists of a sun gear, a ring gear and two or three planet gears that mesh with the sun and ring gears.

A planetary unit can be used to give six different speeds. Two are faster and two are slower than the input speed. The other two speeds run in reverse; one faster and one slower than the input speed.

Some tractors have an automatic gearbox with one or more planetary units. A range of speeds is obtained by using hydraulic clutch packs, which prevent rotation of the sun gear or the ring gear or the planet gear carrier.

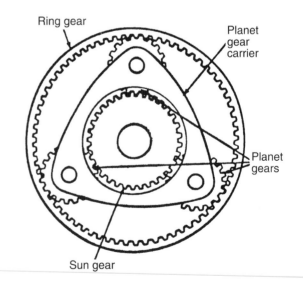

Ring gear

Planet gear carrier

Planet gears

Sun gear

Figure 4.11   An epicyclic unit. (John Deere)

The planet gear carrier is the stationary component in a tractor final drive. The ring gear is an integral part of the rear wheel hub and is driven by the sun gear, on the differential half-shaft through the planet gears. The relative speed of the gears depends on the number of teeth (see page 271) and the shaft carrying the rear wheel (Figure 4.10) will turn at slightly less than half the speed of the output shaft from the differential.

Torque

The turning or twisting force in a shaft is known as torque. When the speed of a driven shaft is reduced, the torque in that shaft is increased. Torque in a tractor transmission system is increased in the gearbox, at the crown wheel and pinion and through the final reduction gears.

The torque available at the flywheel at different engine speeds is sometimes explained in tractor sales literature. Maximum engine torque, which eventually turns the driving wheels, occurs somewhere between half and three-quarters of maximum engine speed. This arrangement gives the engine 'torque backup', so that when the load on the tractor is increased, further engine revolutions are available to bring it back to its maximum torque. When a tractor fails to pull a load in one of the higher gears the problem is overcome by changing down to a lower gear. This reduces the output speed from the gearbox but increases the torque.

FOUR-WHEEL DRIVE

Almost all new tractors have a power-driven front axle to provide four-wheel drive for increased pulling power and enable them to work effectively in difficult ground conditions. Earlier four-wheel drive tractors either had four wheels of equal size, which improved traction but reduced the steering lock, or much smaller front wheels, providing easier steering but much less traction than a tractor with equal-sized wheels. Most current models are a compromise between the two earlier designs and have a diameter about 30 per cent less than the rear wheels. Front-wheel drive may be engaged with a lever after depressing the main clutch pedal or, on more recent tractors, a switch on the instrument panel used while the tractor is in motion.

Many four-wheel drive tractors have automatic four-wheel drive engagement when the rear wheel diff-lock is engaged. Some tractors have an automatic over-ride mechanism which disengages four-wheel drive when the brake pedal is depressed or the forward speed exceeds a certain level – usually between 15 and 23 kph (9 and 12 mph), depending on the model.

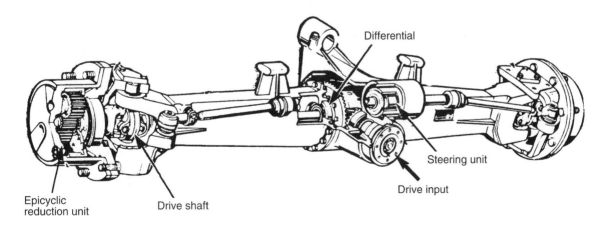

Figure 4.12   The line of drive to the front wheels of a four-wheel drive tractor. (John Deere)

Drive to the front wheels is transmitted forward from a multi-plate clutch in the transmission by an enclosed shaft to a housing on the front axle which contains a crown wheel and pinion and differential. The differential output shafts transmit power to the front wheels. These shafts have a flexible drive to facilitate steering and reduction gears, usually in the form of an epicyclic unit, to give the correct shaft speed for the front wheels.

Front-wheel drive controls vary with different makes and models of tractor. Some have a diff-lock in the front axle that engages and disengages automatically if one front wheel

Plate 4.4   Cutting and conditioning grass for silage with front- and rear-mounted power take-off driven mower conditioners. (Kuhn)

Plate 4.5   Cutting grass for silage with a front-mounted, power take-off driven mower. (Kverneland Vicon)

turns a pre-set amount faster than the other. An alternative system has a limited slip differential which facilitates cornering and equalises the drive to both front wheels when in work.

POWER TAKE-OFF

Standard power take-off speeds are 540 and 1,000 rpm. The 540 rpm shaft is 35 mm in diameter and has six splines. The 1,000 rpm shaft, used to drive machines with a high power requirement, has twenty-one splines or serrations on a 44 mm diameter shaft. Many tractors have only a 540 rpm power take-off, while others have both 540 and 1,000 rpm power shaft speeds selected with a lever or a switch. The six and twenty-one spline shafts are interchangeable on some tractors. When using a fertiliser broadcaster or other implement with a low power requirement, fuel can be saved by using the 1,000 rpm power shaft at a reduced engine speed to run the shaft at 540 rpm.

The increased use of front-mounted, power-driven implements, including power harrows and mowers, has resulted in many tractors having a 540/1,000 rpm front power take-off shaft and front hydraulic linkage. A switch engages drive through a multi-plate clutch and a safety mechanism prevents the power shaft turning after the engine has stopped.

The rear power take-off shaft rotates in a clockwise direction when viewed from the rear. Drive may be engaged in various ways. A switch-operated electro-hydraulic multi-plate clutch is used to engage and disengage the power take-off drive on most tractors. Older tractors have a lever to engage the drive after depressing the clutch. More recent models have a hydraulic multi-plate clutch in the transmission housing, engaged with a separate lever, without using the transmission clutch. Some power take-off shafts have a stop button that stops the shaft instantly and drive cannot be re-started until a security switch is used.

Ground speed power take-off, available on some tractors, has a shaft speed related to the forward speed of the tractor. When the tractor is reversed with the ground speed power shaft engaged, the shaft also turns in reverse. Ground speed power take-off is used when the speed of the drive shaft to the machine must increase or decrease at a controlled rate to match changes in tractor forward speed. A trailer with a power-driven axle is an example of its use.

There are a number of legal requirements concerning the guarding of power take-off shafts. A guard must cover the top and sides of the tractor's splined shaft and it must be capable of supporting a weight of at least 114 kg. When the power take-off is not in use this guard may be removed, but must be replaced with a dust cap of equal strength. The drive shaft from the tractor to the first fixed bearing on the machine must be covered on all sides and there must be a guard over the input power take-off shaft on the machine. It is important to keep all power shaft guards in good condition.

BRAKES

All modern tractors have disc brakes, but some older models still in use have expanding shoe brakes. Brakes are always more efficient when they are mounted on a fast-moving shaft because, with less torque in a high-speed shaft, it requires less effort to slow it down or bring it to a stop. It is usual to have the brakes on the output shaft from the differential before the final reduction gears reduce the shaft speed and increase the torque. On earlier models without final reduction gears, the brakes are in the rear wheel hubs. These brakes are less efficient as they are on the slow speed and high torque half-shafts.

Disc Brakes

A disc brake has a much larger braking surface area than an expanding shoe brake of equivalent size. Most tractor disc brakes are immersed in oil in the transmission housing, but some older tractors have dry disc brakes.

A dry disc brake has one, two or three disc brakes, with a friction lining on each face,

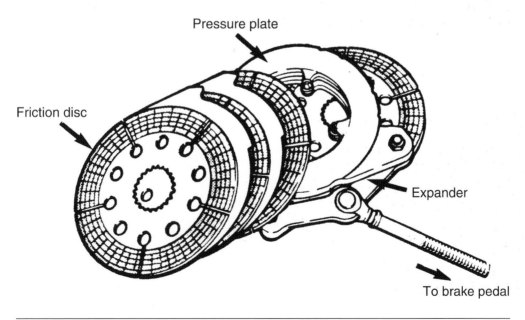

Figure 4.13   A wet disc brake with four friction discs. The expander mechanism moves the two halves of the expander plate outwards to grip the rotating discs and slow or stop the tractor.

which rotate between stationary expander plates. Machined faces on the inside of the brake housing provide outer stationary plates. The discs have internally splined hubs and are attached to the splined drive shafts from the differential to the rear wheels. The two faces of each expander plate are held together by springs and have steel balls in tapered grooves between the two halves of the expander.

When the brake pedal is depressed, the brake linkage causes the two halves of the expander plate or plates to contra-rotate through a few degrees, and the steel balls in their tapered grooves force the expander plates apart. The expander plates grip the rotating friction discs to slow the tractor or stop it completely. When the pedal is released, the springs bring the two halves of the expander plates together again and the wheels are free to turn.

Most tractors have self-adjusting disc brakes, but some require regular adjustment to compensate for wear. The usual method of adjustment is to turn an adjuster nut on each pedal linkage to bring the pedals up to the correct height above the cab floor. The latch, used to lock the pedals together when driving on the road, must be disconnected before adjusting the brakes. Brake adjustment varies with different models of tractor and it is important to refer to the instruction book before carrying out this task for the first time.

In the same way that quiet cabs have brought about hydraulic clutch control, most tractors also have hydraulically operated brakes. Each brake pedal operates a master cylinder, which in turn applies one brake with a small ram cylinder on the disc brake assembly. A balancing valve makes sure that both brake cylinders work at the same pressure when the pedals are locked together.

Oil-immersed disc brakes are usually located in the transmission housing with the brakes mounted on high-speed shafts to achieve maximum braking effect. They may be actuated hydraulically or mechanically and

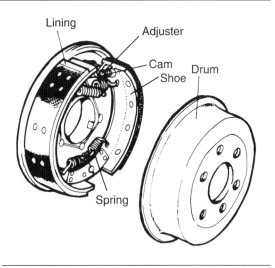

Figure 4.14   An expanding shoe brake.

have between two and four discs in each brake housing.

Wet brakes are more efficient because the oil cools the disc linings and reduces brake fade caused by overheating. The oil also cleans the linings to prevent sticking or binding, a problem with dry brakes caused by dust as the linings wear away. Tractor transmissions with wet brakes may require an oil with special additives to protect the friction faces.

Some four-wheel drive tractors only have oil-immersed disc brakes on the rear wheels, but most have disc brakes on all four wheels. Tractors with automatic front-wheel drive engagement has engine braking on all four wheels when the brake pedals are applied when driving in the higher end of the gear range.

Expanding Shoe Brakes

An expanding shoe brake consists of a pair of brake shoes, both fitted with a friction lining attached to a back plate inside a rotating brake drum. Springs hold the shoes together, and when the brake pedal is pushed down a cam moves the shoes outwards against the drum. The brake drum is secured to the driving shaft and rotates with the rear wheel. When the brake is applied, the pedal linkage turns the cam and the brake shoes are forced outwards. The linings on the brake shoe grip the rotating brake drum, and this slows or stops the tractor. When the pedal is released the springs pull the shoes away from the brake drum.

The brake shoes should be set very close to the drum. They are adjusted by changing the length of the rod between the brake pedal and the brake shoe cam. When the pedals are locked together for driving on the road, it is possible that only one brake will come on if there is uneven wear on the linings. The length of the brake pedal rods must be adjusted (balanced) to ensure that brakes are applied evenly when the pedals are locked together. Unbalanced brakes are very dangerous, especially when travelling at high-speed.

The Parking Brake

All tractors have a parking brake. Older models have some form of locking device used to secure the brake pedals after they have been pushed down with the foot. Some tractors with a mechanical gearbox are parked with either an oil-immersed band brake or a multi-disc transmission brake applied with a hand lever.

Tractors with an automatic or semi-automatic transmission have a built-in parking brake that automatically locks up the transmission when it is in neutral.

Trailer Brakes

Hydraulic trailer brakes, linked to the tractor braking system and operated by the tractor brake pedals, are required for the safe haulage of heavy farm trailers on the highway. A coupling is used to connect the trailer brakes to the tractor and a priority valve arrangement ensures oil is always available when required. Tractors used for high-speed road haulage must have an efficient air braking system for trailers.

Some older trailers have rod or cable operated parking or over-run brakes. A hand lever on the trailer drawbar is used to apply the parking brake. Over-run brakes are linked to the tractor drawbar and the trailer brakes are automatically applied when the tractor slows down.

LUBRICATION

The lubrication of the transmission and rear axle varies with different models of tractor. Change-on-the-move gearboxes are usually pressure lubricated and require special grades of oil. It is also important to use the correct type and grade of oil for wet brakes, wet clutches and for hydraulic systems. As well as lubricating the transmission and hydraulic system, the oil also has a cooling function. Oil is less efficient at high working temperatures, and to overcome this problem many high power tractors have an oil-cooling radiator (Plate 2.5) as well as the engine-cooling radiator at the front of the tractor.

Transmission oils have various additives, including:

- *Anti-wear* – an additive to protect against high pressures between gear teeth.
- *Anti-oxidant* – combats the breakdown of oil (oxidation) which tends to thicken the oil.
- *Brakes* – a special additive is required for transmissions with the brakes running in oil. This overcomes the problems of the brakes juddering or 'squawking' when they are applied.
- *Anti-foam* – reduces the formation of air bubbles in the oil, especially in relation to the use of hydraulic equipment.
- *Anti-corrosion* – as with other lubricants, this gives the oil the ability to combat corrosion, especially from water caused by condensation.

Regular oil level checks and changes should be made in accordance with the operator's handbook. It is particularly important to check transmission oil levels when the hydraulic services are in frequent use with tipping trailer and other auxiliary rams.

CHAPTER 5

Tractor Hydraulic Systems

The hydraulic system provides a means of raising and lowering implements into and out of work and supplies oil to external hydraulic rams and hydraulic motors on trailed or mounted implements. Most implements are carried on the rear three-point linkage, but with the widespread use of high horsepower tractors on the farm, an increasing number of them now have front hydraulic linkage and power take-off. This enables the farmer to make full use of available engine power and carry out two or more field operations at the same time. Examples include a front-mounted furrow press on the front linkage and a rear-mounted power harrow with a drill on the rear linkage. Front and rear mowers are another common combination, with either one or two offset rear units and a central one on the front of the tractor.

Plate 5.1   The hydraulic system provides the power to lift and carry implements at the front and rear of the tractor and supply oil to external rams and hydraulic motors on attached equipment. (Kuhn)

The tractor hydraulic pump also supplies oil at high pressure to hydraulic motors which are used to drive various machines including hedge cutters, fertiliser spreaders and root harvesters. The external rams on tipping trailers, front-mounted tractor loaders and other implements are also supplied with oil from the main hydraulic pump.

PRINCIPLES OF HYDRAULICS

Liquids can be used to transmit power from one point to another. In a simple hydraulic system, a pump supplies oil at high pressure to a ram cylinder. The oil forces the piston (ram) along in the cylinder and the movement of the cylinder is used to lift a load. A hand-operated hydraulic car jack has a hand pump which forces oil against a piston in the cylinder. The piston moves upwards and lifts the vehicle.

Figure 5.1 illustrates a very simple hydraulic lift system. Oil is pumped through the control valve into the ram cylinder. When the control valve is in the '*lift*' position, oil forces the piston along the cylinder. The piston is connected to the cross shaft by a very simple connecting rod. Movement of the piston rotates the cross shaft and the lift rods raise the lower link arms.

When the control valve is moved to the '*lower*' position, the oil is released from the ram cylinder back to the reservoir, which is usually the tractor transmission housing. The implement falls under

its own weight, forcing the oil from the ram cylinder. A relief valve allows the oil to return to the reservoir if the system is overloaded.

All modern tractors have a live hydraulic system with automatic control of working depth and height. However, depth wheels are normally used with semi-mounted and large mounted implements. Some older tractors still in use, especially in livestock farming areas, have a hydraulic system that requires mounted implements to have one or two wheels to control working height or depth.

A live hydraulic system continues to operate when the clutch is depressed, but a few older tractors still in use have a single-stage clutch; and when the pedal is depressed to disengage the transmission and stop the tractor it also disengages the drive to the hydraulic pump.

Pumps

The hydraulic pump, either in the transmission housing or mounted on and driven by the engine, provides pressurised oil to the ram to lift mounted implements. It also supplies oil to external rams and to hydraulic motors. The lift ram is usually inside the transmission housing, but some powerful tractors have external lift rams.

Hydraulic pumps may either be of the fixed or variable displacement (output) type. Output from a variable displacement pump varies according to the demand of the hydraulic equipment in use at the time. The type of pump used will depend on the required oil flow and model of tractor. The three main types are gear, multi-piston and axial piston pump. A gear pump, illustrated in Chapter 2 (Figure 2.11), is similar to that used for many engine lubrication systems, but it has a limited output. Still in use in older tractors, the multi-piston pump has two pairs of horizontally opposed pistons reciprocating in small cylinders. Oil drawn from the transmission housing is pressurised in the pump cylinders and supplied, as

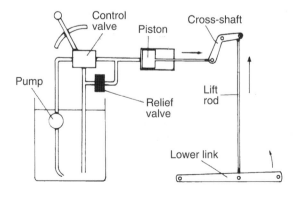

Figure 5.1   A simple hydraulic lift system.

required, via the control valve to ram cylinders and hydraulic motors.

An axial piston pump, which supplies hydraulic oil at high rates of flow and pressure, is used in most medium- and high-powered tractors. The pistons draw oil from the transmission housing or oil reservoir into the pump cylinders through an inlet port. A variable angle plate, or swash plate, on the pump drive shaft operates the pistons, and oil is drawn into the pump cylinders. The variable angle of the swash plate on the pump drive shaft determines the length of the piston strokes and therefore the output of oil from the pump.

Non-return valves retain the oil, which is pressurised in the cylinders before being released to a ram or hydraulic motor.

Relief Valve

This protects the pump from damage through overload by limiting the maximum oil pressure in the system. The oil pressure will increase as the load increases, and when it reaches a pre-set level the relief valve will open and release the pressure in the system. The relief valve will operate if, for example, the driver attempts to lift too much on the three-point linkage or a front-end loader.

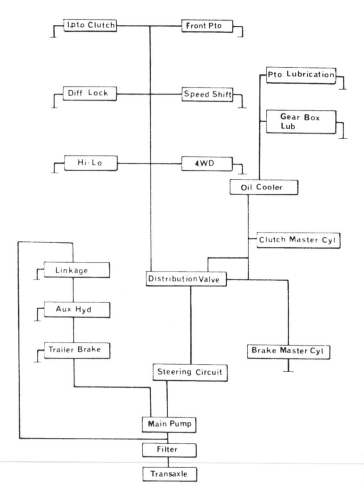

Figure 5.2   Diagram of a typical hydraulic system circuit.

TRACTOR HYDRAULIC CIRCUITS

The hydraulic circuit (Figure 5.2) has high- and low-pressure circuits. The high-pressure circuit supplies oil to the three-point linkage, external ram spool valves and trailer braking system.

The power steering circuit is first in line on the low-pressure circuit and it has a priority oil supply to ensure the tractor can always be steered. A distribution valve also directs low-pressure oil to control the electro-hydraulic power take-off clutch and four-wheel drive clutches. It also engages the diff-lock and actuates the speed shift and high/low changes in the gearbox.

Open and Closed Centre Systems

Older tractors have an open centre hydraulic system, but with the widespread use of power shift transmissions and electro-hydraulic clutches, many tractors now have a closed circuit hydraulic system.

An open centre hydraulic system creates a constant flow of oil, but the pressure varies according to need. The hydraulic pump provides a continuous flow of oil which, when not required for the lift linkage, external ram spool valves or hydraulic motors, is returned to the transmission housing or oil reservoir.

A closed centre hydraulic system has a variable displacement (output) axial piston pump, which only supplies the amount of pressurised oil necessary to operate the hydraulic linkage or other function. At other times, the pump maintains its working pressure but flow is reduced to a minimum level. The advantage of a closed centre hydraulic system with a variable output pump is its ability to supply large volumes of oil for long periods yet save power by reducing oil flow and pressure to a standby level when the system is at rest.

The transmission system oil used for hydraulic circuits is less efficient when it is hot, as it becomes very thin. For this reason some tractors have an oil-cooling radiator in front of the main engine-cooling system radiator.

CONTROL SYSTEMS

Tractors may have levers, switches or buttons to operate the various hydraulic systems and some also have a remote switch at the rear of the tractor for the driver to control the three-point linkage while standing at one side of an implement to aid hitching it.

Draft Control

Draft control is used with implements that work in the soil. Ploughs, cultivators and subsoilers are examples. The draft control system maintains a constant load (draft) on the tractor. In an ideal working situation, with the soil type and condition the same all over the field, the load on the tractor and therefore the working depth would be constant. In practice, of course, this does not happen.

The tractor driver can use draft control to set a plough at the required depth and the tractor will then have a certain load to pull at this ploughing depth. Draft control will maintain this load at a constant level and, at the same time, give an acceptable standard of depth control. Heavy mounted implements, especially ploughs, often have a wheel, or wheels, to help maintain the required depth.

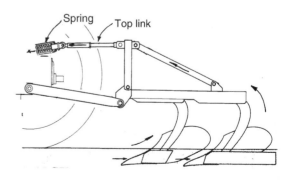

Figure 5.3   How top link sensing works.

Top Link Sensing

A mounted implement is carried on and pulled by the lower links. The top link stops the implement tipping backwards or forwards. When a mounted soil-working implement is in work, a compression force acts along the top link towards the tractor. This force becomes greater as the working depth increases and draft control uses the force in the top link to maintain a constant load on the tractor. There is a linkage inside the transmission housing connecting the top link to the hydraulic control valve. When the implement drops below the required working depth, the strength of the force acting through the top link is increased and the connecting linkage moves the control valve slightly and the implement is raised to its correct working depth.

There is a heavy spring at the tractor end of the top link. This is sometimes visible, but many tractors have the spring inside the transmission housing. The spring, which is connected to the control valve, must be compressed by the forces acting through the top link before any correction to the working depth is made. The spring also marginally pushes the top link in the opposite direction when the force from the implement is reduced. This will happen if the implement is working above the required depth. The spring moves the control valve linkage in the opposite direction and the implement returns to its correct depth.

In very uneven soil conditions there would be too many depth corrections. To overcome this, some tractors have a response or damping control with a range of settings from fast to slow. When depth control operation is sluggish, the response control is set in a fast position. A slow response setting is used when a faster setting is too sensitive and the implement makes too many corrections to its working depth.

The draft control system described above only responds to changes in the compression forces acting through the top link. The weight of an implement working at a very shallow depth can create a tension force in the top link that is greater than the compression force produced by the implement working in the soil.

Some tractors have a double-acting top link which transmits both compression and tension force signals to the hydraulic control valve, and maintains a constant draft (or depth). The top link will be in tension when shallow ploughing. There will be an increase in the tension forces in the top link if the tractor front wheels drop into a furrow which reduces the ploughing depth. With increased tension in the top link, the control valve automatically returns the plough to its pre-set depth.

The top link is always in tension when transporting mounted implements. To prevent damage to the hydraulic system, especially when moving across rough ground, a mechanism cuts out the effect of tension forces in the top link when the implement is fully raised.

Lower Link Sensing

Lower link sensing relies on the fact that as an implement goes deeper, the draft and therefore the tension force in the lower links is increased and this force can be used, through a torsion bar, to operate draft control. Torsion is defined as the twisting force in a shaft. Figure 5.4 shows that when an additional load is placed on the lower lift arms, the torsion bar twists slightly and sends a signal to the control valve which reduces working depth by a small amount, thereby reducing the load on the tractor. When the load on the lower links is reduced, the torsion bar acts in the opposite direction and the control valve returns the implement to its correct depth.

Electronic sensing pins provide an alternative method of draft control. Load sensing pins are used to attach the lower link arms to the tractor, and variations in loading are measured electronically and signals relayed to the control valve to raise or lower the implement.

Constant implement draft (depth) can also be maintained by using the changes which occur

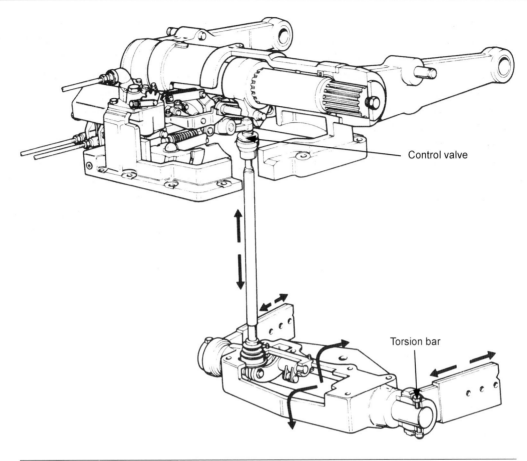

Figure 5.4   Lower link sensing. Increased loading on the torsion bar causes the lower links to raise the implement a small amount. A reduced load returns the implement to its pre-set working depth. (New Holland)

in the torque (twisting force) in the transmission system. Some older tractors have this system, with a coupling, linked to the hydraulic control valve, which reacts to changes in torque in the output shaft from the gearbox. Signals are relayed to the control valve when changes are detected in shaft torque, and the implement is raised or lowered according to need.

Semi-mounted implements can be used to advantage with this system. It monitors variations in draft when ploughing or cultivating depth or soil texture change. The control unit operates a hydraulic spool valve to increase or reduce the working depth of a mounted implement. When using a semi-mounted plough, a hydraulic spool valve linked to a ram on the rear wheel which controls depth of the rear bodies works in combination with the hydraulic lift arms to vary the depth of the front bodies and keep the plough running level.

Electronic Linkage Control (ELC) Systems

The electronic system used for draft control on some tractors also enables the driver to vary the rate of implement drop and to control the maximum height the implement is raised when out of work. It can also lock out the hydraulic system, making it safe when the engine is stopped.

Many modern tractors have a further refinement which gives more accurate control of draft in the hydraulic linkage. This system is pre-set to allow a certain percentage of wheelslip; the normal acceptable level of wheelslip is about 15 per cent. When it exceeds the pre-set level the implement is automatically lifted slightly. As the implement is lifted, the wheelslip is reduced because extra implement weight is transferred on to the tractor rear wheels, which improves traction. The implement returns to the pre-set depth immediately the excess wheelspin has been eliminated. This process may be repeated several times each minute when working in difficult soil conditions.

Without this implement control system, the driver could obtain a similar effect by lifting the implement slightly with the draft control lever when wheel spin occurs. However, fast and accurate manual control would be difficult to achieve and, as sometimes can happen in difficult conditions, the tractor can get stuck if the driver is not quick enough with the control lever. Modern electronic systems have overcome this problem.

Position Control

This system is used for mounted implements such as sprayers, fertiliser spreaders and mowers, which work above ground level. The position control lever is used to set the required working height of the implement on the three-point linkage and any forces acting through the top link will have no effect on the control valve.

The hydraulic system on many tractors can be used with a mix of position and draft control. With the control set near the position mode there will be little reaction to changes in draft, and implement depth will remain reasonably even in varying soil conditions. With the mix predominantly in draft mode there will be more reaction to changes in draft sensitivity, and implement depth will vary from the pre-set depth as soil conditions change within a field from hard to soft or light to heavy.

Flow Control

The rate of oil flow to or from the ram cylinder can be varied on some tractors with a hand-operated flow control valve which is used to control either

the rate of lift or drop depending on model of tractor. On modern tractors, especially those with electronically controlled hydraulic systems, the flow rate and speed of response for the rear and front linkages and the spool valve hydraulic outlets can be altered from the cab.

AUXILIARY SERVICES

The tractor hydraulic system is required to supply oil to other hydraulic services in addition to the three-point linkage. Many implements have external hydraulic rams and hydraulic motors, both of which rely on the tractor to provide a continuous flow of oil.

One of the most common uses for these auxiliary services is to supply oil to the ram on a tipping trailer. This is a single-acting ram which receives oil from the tractor hydraulic system

Plate 5.2 Spool valve couplers allow the tractor to hydraulically-operate auxiliary rams and hydraulic motors on trailed or mounted equipment. The protective cover plates are colour-coded for ease of identification.

Plate 5.3   The tractor hydraulic system supplies oil to the external rams and hydraulic motor on this flail hedge cutter.

when a lever or switch is used to operate the spool valve and tip the trailer. The weight of the trailer body returns the oil from the ram to the tractor transmission system when it is lowered.

On many farms, wide implements which have to be folded manually before taking them on the road are now rarely used. At least one auxiliary ram, used to fold it for transport, will be found on almost every farm machine. While a single-acting ram relies on gravity to return it to the rest position, a double-acting ram provides force in both directions. A tractor loader is an example of a machine with single-acting rams that relies on the weight of the booms to lower it to the ground. Some modern loaders have double-acting rams, which lower it to the ground, and for example also allow the loader to be used to dig into the ground or grade a level surface.

Switch- or lever-controlled spool valves are used to operate single- and double-acting rams. The hoses have quick release connectors that are plugged into hydraulic tapping points – spool valves – at the back of the tractor. Tractors may have as many as eight spool valves, allowing four separate double-acting rams to be connected to the hydraulic system.

The auxiliary hydraulic services also provide a flow of oil for hydraulic motors. Two connections are used, one for flow and the other for the oil to return to the transmission system. Hydraulic motors eliminate the need for expensive implement drive shafts, which need to be guarded, and the oil flow to the motor can be regulated to vary the speed, in forward or reverse, at the touch of a switch or lever. A hydraulic motor is, for example, used to drive the flail rotor on a hedge cutter.

HYDRAULIC LINKAGES

The lower links are connected to the ram cross shaft by the lift rods. The right-hand lift rod has a levelling box which is used to adjust the height of the right-hand lower lift arm and tilt the implement, such as a plough, to one side when required.

The top link, usually adjustable in length, is used to vary the pitch (angle) of some mounted implements including ploughs, cultivators and fertiliser broadcasters. Shortening the top link will tilt the top of the implement closer to the tractor and can help to penetrate hard ground when, for example, using a cultivator.

Stabiliser bars are used to hold a mounted implement rigid on the tractor. Some tractors have telescopic stabilisers that can be set in a fixed position or adjusted to allow some sideways movement of the implement.

External check chains, used on some tractors, serve the same purpose as stabiliser bars and, when necessary, can be used to limit sideways movement of the lower lift arms.

Internal check chains allow free sideways movement of the lower lift arms but prevent them hitting the tyres when an implement swings sideways on the linkage.

The linkage hitch pins on mounted and semi-mounted implements are made in four sizes. They vary in diameter and length and the measurement between the three hitch point positions on the implement is also different:

• *Category One* is the smallest size and is used for compact tractors.

- *Category Two* is the standard size and is used for medium horsepower tractors. Many of these tractors have dual linkage with both categories one and two. This is achieved by changing the hitch balls on the implement ends of the lower links and top link.
- *Category Three* is a heavy-duty linkage on high horsepower tractors and used with very heavy implements.
- *Category Four* is found on some of the most powerful tractors in the 260 kW (350 hp) plus power range, mainly four-wheel drive rigid and articulated models and rubber tracked crawlers.

Most tractors now have quick-attach lower links and top links, but this may not be the case with compact tractors and older models with category one and two hydraulic linkages. After reversing as squarely as possible up to the implement, the easiest, most efficient and safest way to attach an implement to these tractors is to start with the lower left-hand link. Then attach the right-hand lower link, which can be adjusted for height with the right-hand lift rod levelling box, and finally attach the top link, which may be adjustable in length. The linkage should be removed in the reverse order.

Quick-attach linkage arms with hook ends and self-locking latches instead of ball ends are used on most modern tractors. The telescopic ends allow for inaccuracies when reversing up to the implement, which has ball-shaped bushes with conical guides on the hitch pins. The implement is attached by reversing up to it with the lower link arm hooks positioned under the balls. When the lift arms are raised, the hooks engage with the balls and are locked together. The implement end of the top link is then connected in a similar way, but the hook end is clamped over the top of the ball rather than coupling it from underneath. Raising the linkage to retract and lock the telescopic ends in the lift arms completes the hitching procedure.

Uncoupling from the tractor seat is achieved by releasing the three-point linkage hook end locks, then the top link is lifted clear and the lower lift arms are dropped to their lowest point.

Front Linkage

Front hydraulic three-point linkages are in widespread use, especially on arable farms where both front- and rear-mounted implements are used to save passes when cultivating

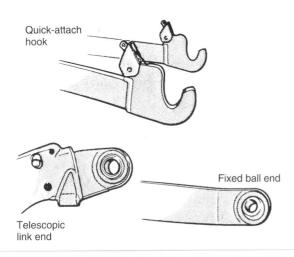

Figure 5.5   Alternative lower link ball ends which slide on to the implement pins. (New Holland)

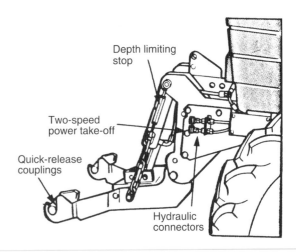

Figure 5.6   Front linkage and power take-off shaft. (New Holland)

Plate 5.4   Tractor front power take-off and three-point linkage with quick release lower link arm ends.

and drilling, or where they have been adopted as a quick method of adding or removing ballast weights. For example, some farmers use a rear-mounted plough and a front-mounted furrow press. Another example is the use of a front-mounted sprayer tank or hopper with a rear-mounted sprayer or seed drill.

Most front-mounted three-point linkage attachments have lift arms with fixed, semi-floating or floating working positions that fold vertically when not in use. Electronic front linkage controls can be used on tractors equipped with electrically operated spool valves. Many front linkages are combined with a dual speed power take-off for front-mounted implements. Drive is engaged with an electro-hydraulic, oil-immersed multi-plate clutch.

Typical rear linkage lift capacities, depending on make and model of tractor, range from about 3 tonnes with a 60 kW (80 hp) tractor to 5 tonnes for an 80 kW (110 hp) model and upwards of 10 tonnes for a 220 kW (300 hp) tractor. Typical front linkage lift capacities vary from about 2 tonnes with a 50 kW (70 hp) tractor up to 8 tonnes for a 220 kW (300 hp) model.

Care and Use of the Hydraulic System

1. Do *not* lubricate the balls in the lower lift arms – they will collect grit, resulting in rapid wear.
2. Never turn a sharp corner with a mounted implement still in the ground – this puts an unnecessary strain on the linkage. A trailed implement, depending on its design, may be more forgiving, but turning with a soil-engaging implement still in the ground is also inadvisable as damage may occur.
3. Always lower a mounted implement to the ground before leaving a parked tractor.
4. Check that the radiator (see Plate 2.5), used on some high horsepower tractors to cool the oil used in the hydraulic system, is not blocked with dust, chaff, etc.
5. Check the transmission oil level – or separate hydraulic system oil reservoir – especially if there has been a loss of oil when using external rams.
6. Never tow anything from the top link position, as the tractor may tip rearwards.

CHAPTER 6

Wheels, Tracks and Steering

Plate 6.1   Four-wheel drive tractors do most of the work on modern-day arable farms. (McCormick)

Plate 6.2   This two-wheel drive tractor with steel wheel rims has front and rear tyres with a grassland tread pattern to limit surface damage. (John Deere)

The efficiency of a tractor can be measured by comparing power developed at the engine flywheel with that available at the drawbar or hydraulic linkage to pull an implement. The transmission, hydraulic system and other components absorb engine power while still more power is lost to the rolling resistance of the tyres and to wheelslip. When power available at the drawbar is compared between three tractors with the same engine output – a crawler, a four-wheel drive tractor and a two-wheel drive tractor – the tracked machine will be the most efficient, followed by the four-wheel drive tractor.

WHEELS AND TYRES

Although almost every new tractor bought by British farmers now has four-wheel drive as standard, many small and medium powered two-wheel drive tractors are still in everyday use, and some manufacturers still make such models.

The front wheels of two-wheel drive tractors consist of a small dished steel disc welded to the rim and fitted with a ribbed tyre. The rear wheels have a dished steel centre bolted to lugs on the rims. Alternative positions for attaching the disc to the rim provide the means of varying track width for ploughing, rowcrop work, tramlining, etc. Rear wheels with narrow rims and tyres are used by some farmers for rowcrop work.

Most four-wheel drive tractors have dished disc wheels on both axles, and on many tractors the front wheels are smaller in diameter than the rear wheels. Many high horsepower four-wheel drive models have heavy cast steel rear wheels.

Tyres

The driving wheels of farm tractors have tyres with an open centre tread pattern. They are fitted to the wheel so that, when viewed from the rear, the arrow pattern of the tread points upwards.

Open centre tread tyres have a self-cleaning action. The tyre wall bulges outwards at the 6 o'clock position of rotation and returns to its normal shape at the 12 o'clock position. The changing shape of the tyre wall as the wheel rotates flexes the tyres to give a self-cleaning action that helps to remove the soil trapped between the tread bars.

Soil compaction can be a major concern for arable farmers. The use of very wide flotation tyres can increase traction and reduce ground pressure by distributing the weight of the tractor over a large footprint area. This is a considerable advantage when preparing seedbeds and drilling or spreading fertiliser and spraying crops during their early stages of growth.

The front wheels of two-wheel drive tractors have a high central rib that aids steering and helps to keep the tractor on the correct course in soft or loose soil. Tractors and farm machinery may have either tubed or tubeless tyres.

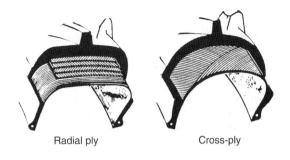

Radial ply   Cross-ply

Figure 6.1   Construction of radial and cross-ply tyres.

Radial and cross-ply tyres differ in construction. The radial tyre is more expensive but has the advantage of improved grip with less soil compaction when compared with a cross-ply tyre. The layers or plies of a radial tyre carcase are placed at right angles to the circular wire bead around the rim. Additional layers on top of the radial plies under the tread (Figure 6.1) give added strength to the tyre carcase. Radial tyres noticeably bulge at ground level when inflated to their minimal permissible working pressure. This gives a larger footprint area and reduces soil compaction.

Radial tyres with open centre tread are now standard equipment on the driving wheels of farm tractors. Until more recent times, cross-ply tyres were fitted on the front and rear wheels and may still be found on some older tractors and farm implements. The carcase of a cross-ply tractor tyre has alternate layers, or plies, which cross over at opposing angles between the wire beads around the rim of the tyre. They have a less flexible tyre wall which gives good resistance to wall damage and high impact loads. However, the tyre wall will be damaged if it is used for long periods with the inflation pressure below the recommended setting for the work in hand.

Tyre Markings

The markings on the side wall of a tractor tyre include the rim section width and rim diameter dimensions and the letter 'R' indicates radial

Plate 6.3   The low ground pressure flotation tyres on this tractor spread weight to reduce soil compaction.

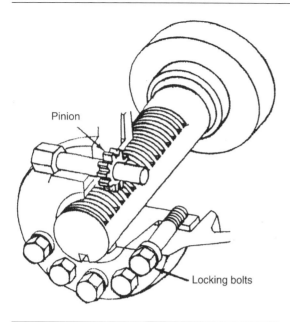

Figure 6.2　A special spanner is supplied for adjusting the track setting on tractors equipped with a rack and pinion track adjustment system.

construction. Other letters and figures refer to the load index and speed code.

Tyre size　Along with other information on modern tractor tyres, the size indicated on their sidewalls uses metric measurements, although the rim diameter they are intended to fit is still given in inches. Older tyres, and those manufactured in North America, may still have imperial measurements with parts of an inch shown as a decimal. Typical examples of agricultural tyre sizing are:

- 710/75 R42 is an example of a modern metric size tyre for the rear wheels of a high horsepower tractor. The first number indicates the section width of the tyre in mm. The second is a measure of the tyre profile or sidewall height and this figure is a percentage of the tyre width.
- 12.4 R 36 is an example of an old North American tractor tyre sizing. Indicating that it is a 12.4 in. wide radial tyre for a 36 in. diameter rim. Some older tyres still in use are marked 12.4 / 11–36 R. They fit a rim 11 in. wide and

36 in. diameter but changes in wheel rim design altered the effective tyre width to 12.4 in. and the 11–36 tyre was re-sized as 12.4 / 11–36.
- 7.50–16 is an example of a cross-ply tyre for tractor front wheels and farm implements. It fits a wheel rim 7.5 in. wide and 16 in. diameter and a 9.00–28 cross-ply tyre is for a 28 in. diameter rear wheel with a 9.0 in. wide rim.

Load index provides information about the use of the tyre. The load index is a three number code, which bears no relationship to the load capacity of the tyre, and this information is used by design engineers. A load index of 90, for example, indicates that the maximum load the tyre can carry is 600 kg. A tyre with a load index of 140 has a maximum load capacity of 2,500 kg and one with a load index of 190 has a maximum loading of 10,600 kg.

Speed Symbol or Code indicates the maximum safe speed for the tyre. Agricultural tyres with a speed symbol of A6, A7 and A8 are limited to 30, 35 and 40 kph (18, 22 and 25 mph) respectively.

The Aspect Ratio, which refers to the cross-section of the tyre, is found by dividing the tyre section height by the tyre width. For example, a tyre with a width of 300 mm (12 in.) and a height of 225 mm (9 in.) will have the following aspect ratio:

$$\text{Aspect ratio} = \frac{225}{300} = 0.75 = 75 \text{ per cent.}$$

This indicates that the tyre section height is 75 per cent of the tyre section width, and a high aspect ratio indicates that the tyre has a very rounded section.

Tyre Care

The information provided by the load index and speed code on the tyre wall is used by the manufacturer to recommend tyre inflation pressures for each model of tractor. The instruction manual gives a range of tyre inflation pressures for different working conditions.

For maximum tyre life, it is important to maintain the correct inflation pressure for the type of work in progress. When tractors are used for continuous roadwork, the pressure should be increased to prevent excessive tread wear and fuel use. Tractor manufacturers also advise increasing front tyre pressure when using a front-end loader.

In pre-metrication days, tyre pressures were measured in pounds per square inch, and this measurement is still in common use. Tractor instruction books may give tyre inflation pressures in psi, kg/cm^2 or, more commonly, the metric 'bar' measurement. One bar is equivalent to 14.7 psi; this is atmospheric pressure, so a tyre with a pressure of 1 bar will be about 15 psi (NB: 1 kg/cm^2 is approximately equal to 1 bar).

Tyre inflation pressures depend on the type of tyre, its use and the maximum load the wheel is required to carry. Radial tyre pressure can vary from 0.4 bar (5 psi) for wide flotation tyres, but typical inflation pressure for a standard radial tyre is 1.6 bar (23 psi) for field work and up to 2.4 bar (35 psi) for high-speed tractors. Typical inflation pressures for cross-ply tyres are 1 bar for rear tyres and 2 bar for front tyres.

On some high horsepower and specialist tractors, tyre pressure can be adjusted from the cab. A pressure control and a small compressor unit on the tractor can be used to decrease tyre pressure to achieve greater tread to ground contact area and improved flotation on the field. The pressure is increased again for stable and safe tyre performance at high speeds on the road. Always check the instruction book to find the correct pressures when inflating the tyres on your tractor, taking into account the operating weight of the tractor and the work being done.

Correct pressures are important to achieve a long tyre life, which will be shortened if the following points are ignored:

- Avoid skidding and fierce use of the brakes and clutch.
- Wipe away any oil or grease accidentally spilled on a tyre.
- Remove sharp objects picked up by the tyre.
- Always replace the dust cap after inflating a tyre.

Improving Tyre Performance

Tractor tyres drive partly by penetration of the lugs in the soil and partly by adhesion (surface contact). A car tyre transmits drive by adhesion alone. Tyre performance can be improved on a tractor by adding weight to improve both lug penetration and adhesion.

High tyre pressures reduce tractive efficiency. The lowest recommended pressure should be used for normal field work to obtain maximum grip.

Although now rarely used, water ballasting is a cheap way to add weight and improve traction. However, this method is not practical for more powerful four-wheel drive tractors with large tyres. Rear wheel weights are more often used to improve traction, and front weights can be added to counterbalance the weight of a heavy rear-mounted implement. Many high horsepower tractors also have cast steel wheel centres.

Weights Although, when compared with water ballasting, cast steel weights are an expensive method of adding weight, they can be removed very quickly and repairing punctures is less of a problem. Weights can be bolted to the rear wheels, carried on the front three-point linkage or weight bracket, or on the rear hydraulic linkage. Rear wheel weights consist of heavy iron or steel discs, each weighing about 40 kg, with two, three, or more of them bolted to each rear wheel. Weights weighing between 30 and 40 kg can be bolted to the inside of the front dished

Plate 6. 4   Front weights are attached to a tractor to counterbalance the weight of heavy rear-mounted implements.

wheel discs to help counterbalance the weight of a rear-mounted implement.

Front weights attached to the three-point linkage at the front of the tractor or added as 'wafers' to a front-mounted weight carrier are more commonly used to counterbalance the weight of heavy rear-mounted implements. Front weights, which may be made of concrete, a composite material or more commonly of steel, are available in various weights. The amount of weight carried on the front of the tractor, typically between 400 and 2,000 kg, will depend on the size of the tractor and the implement. Rear-mounted weights, used to counterbalance a front-end loader, are carried on the rear three-point linkage.

Water ballast Although now less commonly used, water ballasting is a cheap way to add weight and improve traction. However, nowadays it is much less practical for more powerful tractors with large tyres. It is usual to fill the tube about three-quarters full of water with calcium chloride added to protect it from frost. The chemical, usually 1 kg of chemical for every 5 litres of water, should be added to the water. The resulting chemical reaction will generate heat so the solution must be left to cool before filling the tube. A special water-ballasting

valve is fitted to the tube and with the valve at the 12 o'clock position on the wheel it is used to fill the tube to the correct level. The valve allows air to escape from the tube as the water is added. When the tube is 75 per cent ballasted (three-quarters full) the water will be up to the level of the valve. The tyre is then inflated to its normal working pressure and this is checked with a special tyre gauge intended for checking the air pressure in ballasted tyres. The gauge should be rinsed in clean water after use.

Weight transfer The tractor hydraulic system can be used to improve traction by transferring some of the implement weight when in work on to the tractor. This may either be through a mechanical linkage from the top link to the control valve or by means of an Electronic Linkage Control (ELC) system with sensors on the lower lift arms.

Dual wheels A second pair of rear wheels, sometimes attached with a quick-fit mechanism, serves two purposes. When used for seedbed work, they increase the footprint area of the tractor to improve traction and reduce soil compaction. Dual wheels can also be an advantage when working on soft or marshy land.

Plate 6.5   Depending on the pressures used, the increased contact area provided by dual wheels helps to maximise use of engine power or minimise soil compaction.

High horsepower four-wheel drive tractors can have dual wheels on both axles to make full use of the available engine power.

Rubber tracks provide another way of reducing soil compaction and increasing pulling power, especially on high horsepower tractors. Track belt widths can vary from 460 to 920 mm (18 to 36 in.). Unlike the now rarely used steel track crawler tractors, rubber-tracked models do not have the disadvantage of being unsuitable for use on the public highway. A typical rubber-tracked tractor has a top road speed of about 40 kph (25 mph). Rubber tracks are also used on the front axles of some combine harvesters and self-propelled root harvesting machinery.

Diff-lock The differential lock (see page 49) is another valuable aid to traction, especially when wheel spin is a problem.

WHEEL TRACK SETTINGS

Maximum and minimum front and rear wheel track settings will depend on the size of the tractor. High horsepower models have very heavy wheel centres. The wheel hubs can be moved in or out on the axles to alter the track setting on some models. Others have a fixed track setting, a legal requirement for tractors with a top road speed of 50 kph (30 mph).

Rowcrop wheels are usually adjusted in 100 mm (4 in.) steps from 1.22 or 1.32 m (48 or 52 in.) up to 1.83 or 1.93 m (72 or 78 in.). A rowcrop planted at 500 mm (20 in.) row spacing will require a wheel track setting of 1.9 m.

Tractors in the 60 kW (80 hp)-plus bracket may have a minimum rear wheel setting between 1.43 or 1.53 m (56 and 60 in.) up to a maximum of 2.18 m (86 in.). Front wheels may be adjustable from 1.53 m to 2.33 m (60 to 92 in.) on two-wheel drive models, but the range is slightly less on four-wheel drive tractors. The front-wheel track width on some four-wheel drive tractors cannot be adjusted.

Plate 6.6   Power adjusted rear wheels. Engine power is used to rotate the wheel centre on spiral rails inside the wheel rim to alter the rear track setting.

Changing the position of the wheel centres in relation to the lugs on the rims varies the wheel track setting on most rowcrop tractors. The wheel discs can also be reversed to alter the track setting.

Power adjusted rear wheels (Plate 6.6) are fitted to some of the more powerful tractors. The rim is like a giant nut and the wheel disc is the bolt. The wheel track is adjusted by rotating the disc in the rim using engine power and using a low forward or reverse gear depending on whether the track width is to be increased or decreased. The wheel being adjusted is allowed to turn, while the other is locked with the foot brake. Clamps that hold the disc in position on the rim are removed. A stop is placed in one of a set of holes in one of the rails to obtain the required track setting and once adjusted the clamps are replaced and secured.

Rack and pinion wheel adjustment (Plate 6.7), used on some high horsepower tractors, provides infinite wheel track settings within the minimum and maximum track widths of a particular model. The mechanism consists of a pinion gear in the wheel hub that is in constant mesh with a rack or row of teeth on the rear axle shaft. Bolts are used to clamp the wheel hub on to the axle.

Plate 6.7   A rack and pinion mechanism is used to change the rear wheel track width on this tractor.

Track width is altered by turning the adjusting pinion with a spanner after slackening off the locking bolts and jacking the wheel clear of the ground.

Two-wheel drive tractors have a telescopic front axle. The front track setting is adjusted by removing the axle beam bolts and then sliding the movable section inwards or outwards to the required setting. Very wide track settings are obtained by reversing the wheel discs on their hubs so that they are dished outwards. It will be necessary to adjust the toe-in on most tractors after altering the track width, which is measured at ground level.

Rear tractor wheels are very heavy, especially when they have cast steel wheel centres. Make sure you have the proper equipment, including lifting tackle, and have help available when removing or fitting a rear wheel. Slacken the wheel nuts half a turn before jacking up the tractor and make sure the axle is securely supported with an axle stand or heavy wooden block before removing the wheel. Chock the other wheels with blocks. When replacing a wheel, take care not to damage the threads on the wheel studs. Fit the top nut first. When all the nuts are finger tight, tighten the top nut first and then the remaining ones in diagonally opposite pairs. Carry out the final tightening after lowering the wheel to the ground.

Axle stands should be used to support the front axle, and the rear wheel brakes should be applied before adjusting the wheel track on a two-wheel drive tractor.

TRACKS

Steel-tracked tracklaying or crawler tractors were popular for many years, especially on heavy land farms. An important feature of a

Plate 6.8   The front axle on each side of a two-wheel drive tractor is in two parts and held together with bolts. The front-wheel track setting is adjusted by moving the outer axle sections and wheels closer to or away from the tractor.

Plate 6.9   This rubber-tracked crawler with large rear drive sprockets and smaller front idler wheels has a 290 kW (395 hp) engine and a top speed of about 25 mph (40 kph).

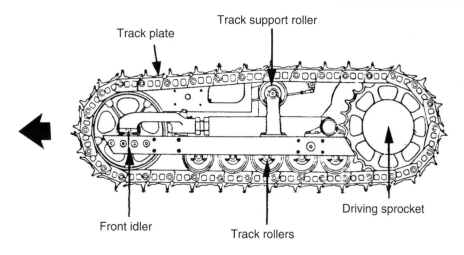

Figure 6.3   The track layout on a steel-tracked crawler tractor. Rubber-tracked crawlers, which can be driven on the road, have largely replaced steel-tracked models.

tracklayer, when compared with a wheeled tractor, is the very low ground pressure exerted by the tracks, which results in reduced soil compaction.

Powerful four-wheel drive tractors and rubber-tracked crawlers have almost completely replaced the slow-speed steel-tracked models that can only be driven on the highway when fitted with track plates to protect the surface of the road, and are consequently slow and noisy.

Conventional steel tracks (Plate 1.8) consist of a series of track plates carried on a very strong endless track chain with rectangular links. The track is carried on a set of rollers mounted in the track frame, with a large driving sprocket at the rear and an idler wheel at the front (Figure 6.3). The bottom rollers hold the track against the ground, and either one or two top rollers support the upper section of the track. The teeth on the driving sprocket engage with the track plate links to pull the track round the frame and propel the tractor. The track is tensioned by adjusting the position of the front idler wheel on the track frame. Various track plate widths are used, but 400 mm (16 in.) is a typical width.

Rubber Tracks

A continuous rubber belt with a cleated track pattern on the outer surface forms the track on rubber-tracked tractors. One design (Plate 6.9) has a large diameter driving wheel on the rear axle with a smaller front idler wheel and a set of track rollers to maintain contact between the track and the ground. Pegs on the inside of the steel wire reinforced tracks hold them in position by running in grooves in the centre of the driving wheels, idler wheels and track rollers. Engine power is transmitted from the transmission system to the track via the rear drive sprocket bars, which engage with lugs on the inner surface of the track belts. Tensioning devices on the idler wheels maintain the tracks at the correct tension.

An earlier track design (Plate 1.7 in Chapter 1) has equal-sized driving wheels at the rear and front idler wheels with a group of idler rollers attached to the track frame. The tracks are also driven by friction between the inner surface of the cleated track belts and the rubber-coated driving wheels.

A different design of rubber-tracked tractor (Plate 6.10), with centre-pivot or articulated steering, has four independent triangular

Plate 6.10   Four independent rubber tracks and articulated centre-pivot steering make this tractor easy to manoeuvre on headlands.

rubber tracks. Like the twin track design described above, the wide tracks are independently driven by lugs on the underside of the track belts that interlock with bars on the driving wheels. The tracks are hydraulically self-adjusted to maintain the correct tension. Both track designs are capable of 25 mph (40 kph) travel on the road.

Wheels or Tracks

There is little difference in the performance at the drawbar of a tractor with four-wheel drive

or rubber tracks. Both rely on having sufficient grip with the ground to pull an implement. This is achieved partly by the adhesion from friction between the tyres or tracks and the soil and partly by the penetration of the lugs in the soil.

Tracks have better traction in loose soil and on steep hillsides and better flotation on wet ground, but tyres do less damage to the crop when working in rowcrops and when turning on the headland. Because the weight of crawler tractors is carried on a large track area, it tends to create less ground pressure, and as a result less soil compaction than wheeled tractors.

STEERING

With the exception of a few models at the lower end of the power range, modern farm tractors have four-wheel drive and hydrostatic steering. However, there are many thousands of tractors with manual or power-assisted steering still in use.

Most wheeled farm tractors have front-wheel steering, but multi-mode steering including front-wheel, rear wheel, four-wheel and crab steering is found on some high-powered tractors and telescopic handlers. Four-wheel steering, where the wheels on each axle turn in opposite directions, gives greater

Four-wheel steering   Crab steering

Figure 6.4   Four-wheel steering, where front and rear wheels turn in opposite directions, and crab steering, where all four wheels turn the same way.

manoeuvrability and a smaller turning circle. Crab steering, where the wheels all turn in the same direction, allows the machine to move forwards and sideways at the same time, for manoeuvring in tight spaces.

Bi-speed turning, on some four-wheel drive rowcrop tractors, drives the front wheels a little faster than the rear wheels when turning at the headland to give the tractor a tighter turning circle.

Articulated or centre-pivot steered tractors have a smaller turning circle than a front-wheel steered tractor of the same size. They have a central pivot with hydraulic rams, actuated by the steering wheel, which changes the direction of travel.

Manual Steering

Some older tractors still in use have manual steering (Figure 6.5). The steering wheel, which is usually connected to a recirculating ball steering gearbox, moves the drop arm forward or rearward through an arc. The drag link connects the drop arm to the steering arm on the front axle, and when it is moved forward or rearward the drag link pushes or pulls the steering arm on both front wheels to change the

direction of travel. The track rod links the two steering arms, and its length has to be adjusted to realign the front wheels after changing the track width.

Another type of manual steering has two-drop arms operated through the steering box. When the steering wheel is turned, one drop arm moves forward and the other moves backwards. The drop arms are connected to the steering arms by separate drag links. There is no track rod. When the steering wheel is turned, the drop arms and the drag links turn both wheels in the same direction.

Power–Assisted Steering

This type of steering has a manual steering linkage with a miniature hydraulic system to help move the steering arms to the left or right. The hydraulic unit consists of a pump, control valve, an oil reservoir and a ram cylinder. The pump is usually driven by the engine fan belt and a linkage from the steering box operates the control valve.

Tractors with power-assisted steering can still be steered with some difficulty if a fault occurs in the hydraulic unit. The ram is double acting. This means that oil can be pumped to

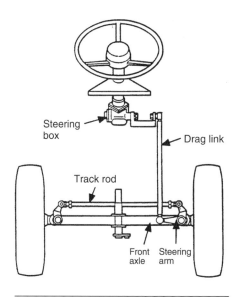

Figure 6.5   A single drop arm steering linkage.

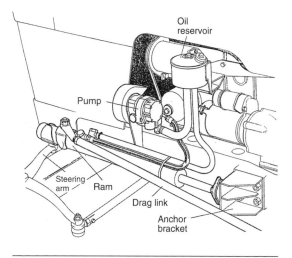

Figure 6.6   The layout of a power-assisted steering system.

either side on the ram piston to give movement in two directions. When turning left, the control valve directs oil to one side of the piston, which turns the front wheels to the left. Oil from the opposite end of the piston is returned to the reservoir.

Hydrostatic Steering

Manoeuvring heavy and powerful tractors with hydrostatic steering and no mechanical linkage between the steering wheel and the front wheels is an easy task. The main components of a hydrostatic steering system (Figure 6.8) are an oil pump, steering ram, oil reservoir and a control valve linked to the steering wheel. Some tractors have an engine-mounted pump supplying oil from a separate reservoir to the steering ram, but most tractors have a steering pump in the transmission housing. A priority valve in the steering circuit supplies oil to the steering ram ahead of any demand from the hydraulic linkage. Some tractors have separate

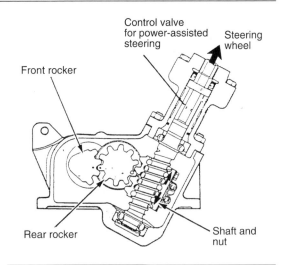

Figure 6.7 Recirculating ball steering box with an oil control valve for power-assisted steering.

pumps in the transmission housing for the steering circuit and for the hydraulic linkage. The control valve, operated by the steering wheel, directs oil to one side of the steering ram

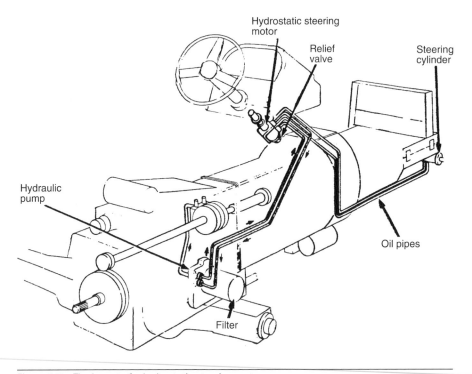

Figure 6.8 The layout of a hydrostatic steering system.

and returns the oil in the opposite side of the ram back to the reservoir or transmission housing, depending on the type of system.

An alternative hydrostatic steering system has a separate hydraulic ram for each front wheel. When the control valve, operated by the steering wheel, directs oil to one of the steering rams, the oil in the opposite ram is returned to the reservoir.

Steering Guidance Systems

There are two main types of electronic steering guidance system. Both are linked to a network of satellites known as the Global Positioning System (GPS) (see page 80).

A basic satellite guidance system uses a cab mounted receiver which records the tractor's position on Earth relative to a series of satellites. A lightbar, or a monitor screen, then provides a parallel tracking guide when, for example, using a sprayer or fertiliser spreader without tramlines. The tractor driver records an initial run, or an A to B line, and then steers according to a series of coloured lights – a light bar – to keep the tractor on a parallel track to avoid overlaps or missed strips.

More expensive steering guidance systems linked to GPS, a third fixed point on Earth and a correction signal, provide greater accuracy down to 2.5 cm (1 in.) between bouts. When equipped with the correct equipment, which acts on the steering rams or steering wheel, the system can be used for hands-free parallel tracking between each pass across the field. The most advanced system also provides hands-free headland turns.

Front Axle Suspension

A smoother ride on uneven ground and consequently higher working speeds are two of the benefits of front axle suspension, which is usually standard equipment on the more powerful four-wheel drive tractors. One design combines a mechanical linkage with hydraulic cylinders and electrical sensors to

Plate 6.11   Hydraulic front axle suspension, front three-point linkage and front power take-off are features of this tractor. (John Deere)

provide a smoother ride and keep the tractor on an even keel. Another design uses two double-acting hydraulic cylinders to control the vertical movement of the front axle with nitrogen accumulators acting as shock absorbers. A third design uses four moving links, a hydraulic ram, hydraulic shock absorbers and an electronic control system, which automatically maintains the front axle at a constant suspension height.

Tracklayer Steering

Two types of mechanical steering system, with a separate lever for each track, are used on steel-tracked crawler tractors.

Clutch and brake steering has a separate hand lever to control a steering clutch on both half-shafts. The tractor is steered by disengaging either the left- or right-hand clutch. Drive continues to the opposite half-shaft and the tractor slews round to take up the required direction. Sharper turns are achieved by applying the brake, either with the same lever or a foot brake, to the inner track.

Controlled epicyclic differential steering is an alternative system. When one of the hand levers is used to apply the brake on one half-shaft, the differential increases the speed of the opposite track and the tractor changes direction. Unlike the differential in a wheeled tractor, where one wheel can be stationary and the other turns at twice normal speed, both half-shafts on a crawler are always under power, but they turn at different speeds when cornering.

Rubber-tracked crawler tractors have a steering wheel connected to a hydraulic control unit which causes the steering differential to speed up one track and slow the other when turning a corner.

CHAPTER 7

Tractor Cabs

Tractor cabs have changed beyond recognition since the days of the first weather cabs and roll frames, and later still, noisy safety cabs. Since the mid-1970s, farm safety regulations have required agricultural tractors to have a quiet cab with a maximum noise level of 86 decibels at the driver's ear. Many modern-day tractor cabs have automatic climate control, air suspension for the driver's seat and the cab itself, digital, electronic and touch screen controls and displays (see Plate 1.14 in Chapter 1), which allow recorded information to be downloaded or even transmitted directly to the farm computer. Headland management systems, which allow a sequence of operations when entering or leaving each pass of field work to be recorded and subsequently carried out automatically at the touch of a button, are another feature found on some of the more powerful tractors.

Plate 7.1   Modern high horsepower tractor cabs have electronic controls, heating and air conditioning, an air suspension seat and a second seat for a passenger. Some have a guidance system linked to GPS satellites. (Massey Ferguson)

CAB SUSPENSION

Some cabs have a pneumatic suspension system which isolates the driver from the bouncing effect of driving over rough and uneven field surfaces. Different levels of suspension can be selected for road haulage and fieldwork. Electronically controlled air cylinders, assisted by dampers, allow the cab to move the cab vertically up or down within controlled limits from the central position to provide an even ride.

Other cabs have electronically controlled hydraulic cylinders linked to nitrogen-filled accumulators that maintain the vertical alignment of the cab and cushion the driver from shocks caused by uneven ground. Flexible rubber mounts absorb vibration, and the hydraulic cylinders have sensors to keep the cab level and some even compensate for varying weights in the cab.

TRACTOR AND ENGINE MANAGEMENT SYSTEMS

An electronic management system in some tractor cabs communicates between the engine and the transmission. This fine-tunes the engine to provide the highest possible power output with optimum fuel efficiency. Suitable maximum and minimum engine speeds for a particular job can be set electronically and retained for future use.

Electronic engine and transmission management systems in the cab combine to provide automatic control of engine speed in relation to forward speed and automatically select the correct gear ratio to match the forward speed when changing up or down within a gear range.

Some tractor transmissions incorporate a cruise control system which, when required, provides set and constant forward speeds. The driver selects the cruise control speed settings – one for working across the field and another when approaching the headland before turning the tractor. The selected cruise speeds are stored by pressing a switch, and every time the switch is used, the tractor will travel at the selected cruise control speed and at the minimum possible engine speed.

Electronics are also at the heart of semi- and fully automatic transmissions which, according to load, can change up or down within a selected or full gear range. Some electronic systems also memorise the driver's choice of engine speeds and gears and repeat the chosen driving style. Some systems have separate field and transport modes. In field mode, and depending on load and required speed, the transmission is programmed to shift up and down through a lower range of gears. For transport, the computer selects suitable road gears and engine speeds with engine braking.

More advanced electronic headland management systems enable the driver carrying out a repetitive task such as ploughing and turning at the headland to control a number of separate functions in any pre-set order or combination with the touch of a switch. The system will also automatically select and maintain the most efficient forward speed, based on the selected throttle setting, for the work in hand; or the most efficient throttle setting for the selected forward speed.

The system can also be programmed to engage and disengage four-wheel drive, differential locks and power take-off, and can also store and repeat a sequence of operational events that use hydraulic spool valves to operate the rams on a mounted or trailed implement.

Steering Guidance Systems

Global Positioning System (GPS) satellites were first used in agriculture for combine harvester yield mapping, which pinpointed areas of fields giving different yields. More recently, GPS has become widely used for guided and automated steering on tractors. Basic systems provide the driver with a satellite-guided manual steering aid, usually a series of coloured lights (a lightbar) to help the driver steer to the centre light and ensure the current pass is parallel to the previous one, minimising misses and overlaps.

A more sophisticated system provides completely automated steering that ensures parallel passes to reduce overlap to a minimum by constantly correcting the steering, either at the steering column or on the axle itself. This relieves the driver from the need to steer the tractor, enabling greater focus on checking implement operation and performance.

Some tractors have a headland management system, linked to satellites, which automates the turning at the headland as well as lifting or lowering the implement, etc. This can be programmed to record and repeat any sequence of actions, including lifting an implement, turning at the headland and setting it back into work. This sequence can, with a few flicks of a switch, be repeated time after time.

A typical example of an automatic electronically controlled headland management sequence when ploughing with a semi-mounted

Plate 7.2   A GPS guidance system automatically steers this tractor parallel to an initially-driven and stored A-B line for accurate matching of subsequent passes. The uncultivated strips of land are exactly the same width as the cultivator, or exact multiples of it. (Massey Ferguson)

reversible plough starts by decreasing the engine speed and forward speed as the tractor nears the headland. The plough is raised on the hydraulic linkage and four-wheel drive is disengaged. After turning the tractor on the headland, the system re-engages four-wheel drive and the plough is lowered back into work. The engine is brought back to its pre-set speed and the tractor returns to the selected ploughing speed.

Electronic Fault Diagnosis

Tractor service schedules are always included in the operator's instruction manual. Many on-board computer terminals store service information and, when required, display it on a screen in the cab. Some tractors also have a port used by a dealer's technician to plug in a laptop computer to diagnose a fault in the engine, transmission or hydraulic system.

ROUTINE TRACTOR MAINTENANCE

There is considerable variation in the regular maintenance requirements of farm tractors, both in service periods and range of tasks required. The operator's manual sets out the correct service schedule and this should be consulted before carrying out routine tractor maintenance.

Remember that the periodic service hours refer to the *engine* hours shown on the proof meter. In many cases, the proof meter registers the hours run by the engine at two-thirds of its maximum speed. So, in a normal day's work, the proof meter will register more hours than the tractor actually works. However, on some tractors the proof meter registers the actual hours worked, i.e. the hours are not related to engine speed.

The maintenance required by a tractor will depend on its specification, and more importantly, its age. Modern high-grade lubricants and cleaner running engines have considerably extended the engine and transmission service intervals compared with older models.

Tractor maintenance schedules and service periods vary with different makes and models and the age of the tractor. The instruction manual should be consulted before carrying out the following tasks :

- Daily servicing should at least include checking the fuel level.
- When working in very dusty conditions, servicing the air cleaner filter element should also be carried out daily. Many tractors have warning lights or gauges in the cab to indicate low fuel and oil levels and when the dry air cleaner filter needs to be serviced.
- Older tractors will need the engine oil level checked daily.
- When using a tractor with an oil-bath air cleaner in very dusty conditions, the oil in the oil bath should be checked; and if there is a build-up of dirt, the oil should be renewed after cleaning the oil bath.
- Depending on working conditions, weekly or 50-hour maintenance tasks should include checking oil levels and servicing the air cleaner.
- Grease points should be lubricated and it is good practice to check that the radiator fins are not clogged with dust, etc.

Engine oil change periods depend on the model of tractor, and this information will be found in the tractor instruction book. Older diesel-engined tractors need an engine oil and filter change usually at intervals of between 250 and 300 hours, but for newer models the

Plate 7.3   Checking engine oil level. (John Deere)

oil change periods are usually between 500 to 600 hours; on some engines it can be up to 750 hours. Other maintenance tasks at this time include replacing the diesel fuel filter element, checking the battery terminals for corrosion and checking fan belt tension (see page 36). Periodic checks include transmission and four-wheel drive oil levels, and brake pedal adjustment on tractors with a mechanical pedal linkage.

The dry air cleaner element and the fresh air fan filter element in the cab should be replaced at intervals of 750 to 1,000 hours. The operator's manual will probably advise changing the oil in the transmission, hydraulic system and four-wheel drive gearboxes at the same time, but check with the tractor manual for the exact service periods.

CHAPTER 8

Small Engines

Small internal combustion engines provide the power for a number of small farm machines including chain saws, brush cutters, standby engines for milking equipment, generators, lawn mowers, etc. Small air-cooled diesel engines are used for some applications, but most small engines on the farm are either two- or four-stroke, air-cooled petrol engines with horizontal crankshafts. Rotary lawn mowers have a vertical crankshaft air-cooled engine.

SINGLE-CYLINDER TWO- AND FOUR-STROKE PETROL ENGINES

The Four-Stroke Petrol Engine

Four-stroke petrol engines have lower working pressures and temperatures than diesel engines. The compression ratio of a petrol engine is about 7:1. This means that the volume in the cylinder at TDC is approximately one-seventh of the volume at BDC compared

Plate 8.1   The pick-up mechanism on this horse manure collector is driven by a small air-cooled petrol engine with a horizontal crankshaft. (Suzuki)

with a compression ratio of 16:1 for a diesel engine (see page 16). Some small engines have side valves, while others have the more efficient overhead valves.

Petrol engines have a carburettor to supply and mix petrol and air at the correct ratio, and

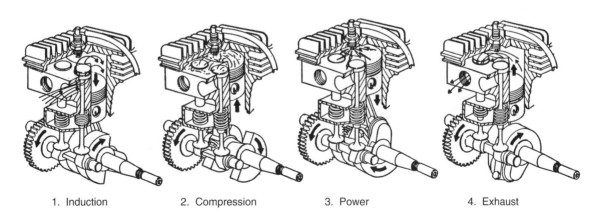

| 1. Induction | 2. Compression | 3. Power | 4. Exhaust |

Figure 8.1   The working principles of a four-stroke petrol engine. (Briggs & Stratton)

an ignition system to provide the spark which ignites the compressed mixture of fuel and air in the cylinder. A four-stroke petrol engine works in this way:

Induction stroke A mixture of fuel and clean air is sucked into the cylinder through the open inlet valve.

Compression stroke Both valves are closed. The rising piston compresses the mixture of fuel and air in the cylinder. Just before the top of the compression stroke, a spark ignites the compressed mixture.

Power stroke The expanding gases in the cylinder drive the piston down, turning the crankshaft.

Exhaust stroke The rising piston forces the waste gases from the cylinder through the open exhaust valve. The cycle starts again with another induction stroke.

The Two-Stroke Petrol Engine

Two-stroke engines are small, lightweight power units suitable for chain saws, brush cutters, etc. Noting the position of the piston in relation to the ports in Figure 8.2 will help the reader to understand the working principles of a two-stroke engine.

The main differences between a two-stroke and a four-stroke engine are that:

- Every other stroke of the piston in a two-stroke engine is a power stroke compared with one in every four in a four-stroke engine.
- The two-stroke petrol engine has no valves. It has openings, called ports, in the cylinder walls which serve the same purpose as

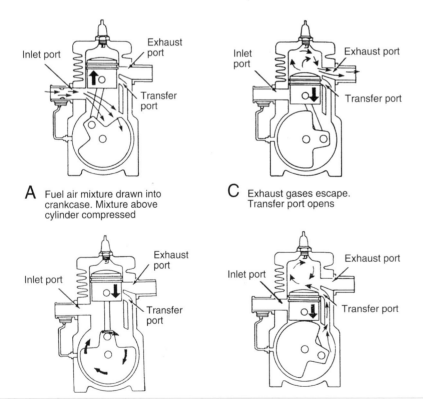

A Fuel air mixture drawn into crankcase. Mixture above cylinder compressed

C Exhaust gases escape. Transfer port opens

B Piston driven down. Mixture lightly compressed in crankcase

D Mixture passes through transfer port into cylinder

Figure 8.2 How a two-stroke petrol engine works. (John Deere)

the valves in a four-stroke engine. There are three ports:

1. The *inlet port* connects the cylinder to the supply of air and fuel from the air cleaner and the carburettor.
2. The *transfer port* connects the crankcase to the combustion space above the piston.
3. The *exhaust port* connects the combustion space above the piston to the exhaust pipe.

• The crankcase is airtight and contains no oil. Lubrication is by oil mist (oil mixed with the petrol). The engine instruction book states the proportions of petrol and oil to be mixed together and this advice should be carefully followed.

First stroke As the piston nears top dead centre (TDC), a mixture of air and fuel (which contains some oil) is drawn into the crankcase. The mixture of fuel and air already above the piston is compressed and then ignited by a spark from the flywheel magneto or the electronic ignition system. Expansion from the resulting heat drives the piston down, turning the crankshaft. Towards the end of this stroke the piston closes the inlet port and the fresh mixture of air and fuel, trapped in the crankcase, is slightly compressed by the descending piston.

As the piston nears bottom dead centre (BDC), it uncovers the exhaust and transfer ports. The exhaust gases are released from the combustion space above the piston through the exhaust port to the exhaust pipe. The slightly compressed air and fuel mixture in the crankcase below the descending piston passes through the open transfer port into the combustion space above the piston.

Some two-stroke engines have a piston with a shaped top which causes the incoming air to swirl, helping to clear the exhaust gases from the cylinder.

Second stroke The piston starts its upward stroke and any remaining exhaust gases are pushed out of the cylinder. The piston then covers the transfer and exhaust ports and the fresh mixture of air and fuel in the combustion space above the piston is compressed. By this point in the two-stroke cycle, the piston has uncovered the inlet port and more mixture of air and fuel is drawn into the crankcase because the rising piston creates a slight vacuum in the crankcase. The compressed mixture above the piston is ignited just before TDC and the cycle starts again.

FUEL SYSTEMS

Fuel flows from the tank, usually through a fine filter, into the carburettor. A tap is provided to turn off the fuel when the engine is stopped. Some small engines have a mechanically driven fuel pump. The purpose of the carburettor is to supply and mix the correct amount of petrol with the large volume of air from the air cleaner as it is drawn towards the combustion chamber. There are various types including float chamber, diaphragm and suction lift carburettors.

Float Chamber Carburettors

This type of carburettor has a fuel reservoir or chamber with a lightweight float which controls fuel level. A small needle valve attached

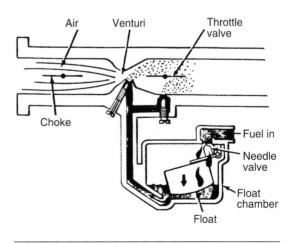

Figure 8.3   A float chamber carburettor. (John Deere)

to the float allows fuel into the float chamber to replace that being used and cuts off the supply when it is full. The needle valve opens and closes many times every minute when the engine is running. A tap in the fuel pipe can be used to cut off the fuel supply to the carburettor when the engine is not running.

Fuel flows from the float chamber through a small pipe to the tube which carries air from the air cleaner to the cylinder. The diameter of this tube is reduced around the main jet, which is called the venturi. Its effect is to speed up the flow of the air in the tube, and this increase in air speed creates partial vacuum around the main jet in the venturi. This causes a fine mist of petrol to be introduced from the main jet into the air stream and mix with the air as it is drawn towards the cylinder by the descending piston.

Float chamber carburettors have a main fuel jet which can be adjusted to give the correct proportions of petrol and air for efficient engine performance. At low engine speeds there is insufficient airflow to create the necessary vacuum to draw fuel from the main jet. In this situation, the idling jet, positioned between the throttle valve and the cylinder, supplies the fuel required to keep the engine running at idling speed.

The slow running or idling speed of a petrol engine is usually adjusted with a throttle stop screw. This prevents the throttle butterfly valve completely shutting off the flow of air and petrol to the engine. The throttle stop screw is usually adjusted with a screwdriver after setting the throttle control in the slow run position until the engine runs smoothly at idling speed.

Some carburettors have another adjusting screw used to set the air/fuel mixture when the engine is at idling speed. The engine may misfire or stop if the mixture is too rich or too weak. Only a fraction of a turn on the adjusting screw is normally needed to achieve smooth running at idling speed. When adjusting the carburettor, turn only one adjusting screw at a time and do so very gradually, taking note of the changes in the way the engine runs.

Some carburettors, particularly those on chain saws, have a main jet (or power jet) which can be adjusted to give maximum power at high engine speeds. Most engines develop maximum power output at less than full revs; this gives a reserve of engine power when the engine is heavily loaded.

Diaphragm Carburettors

A float chamber carburettor will only operate efficiently when the float chamber is in an upright position, making it unsuitable for chain saw engines that require the engine to be tilted sideways when, for example, felling a tree. A diaphragm carburettor, which is not affected by the angle at which the engine has to run, overcomes this problem. Engines with a diaphragm carburettor have many other applications, including brush cutters, strimmers and hand-held hedge trimmers.

One type has a single diaphragm instead of a float chamber and float. In common with an engine with a float chamber carburettor, suction created on the downward stroke of the piston draws fuel from a jet in the venturi and mixes it with the air on its way to the cylinder. This suction pulls the diaphragm upward against spring pressure, which opens the fuel inlet valve and allows more fuel into the chamber.

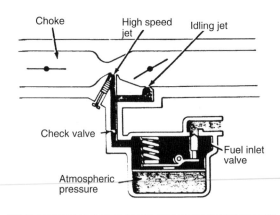

Figure 8.4   A diaphragm carburettor. (John Deere)

During the remaining part of the two-stroke cycle, there is no suction in the venturi and the spring closes the diaphragm, which then remains closed until the next induction stroke.

Another type of diaphragm carburettor has two diaphragms. One draws petrol into the carburettor chamber and the second supplies the required amount of fuel to the air passing through the venturi into the cylinder.

Suction Lift Carburettor

This type of carburettor is usually mounted on top of the fuel tank and has a fuel pipe extending down into the tank. A fine filter screen and a check valve are situated at the lower end of the pipe. The top of the pipe is in the tube which carries the air to the engine. It has an adjusting screw to set the correct fuel and air mixture. A second discharge opening provides fuel to the air stream when the engine is at idling speed.

Vacuum created in the air tube by the descending piston draws petrol from the fuel pipe, which mixes with the air as it passes into the cylinder. With the throttle butterfly closed at idling speed, a smaller amount of fuel is drawn from the idling outlet. A check valve ensures that the fuel pipe is always full when the engine is running.

Another type of suction lift carburettor has a small diaphragm type fuel pump which lifts petrol into a secondary chamber. Vacuum in the air tube draws petrol from this chamber through a small pipe and mixes it with the air passing to the cylinder.

Carburettor Controls

The throttle controls the amount of air/fuel mixture supplied to the engine. The throttle butterfly valve, situated between the carburettor and the engine and controlled with the throttle lever, is used to set the speed of the engine.

The governor over-rides the throttle setting when the engine speed falls away and brings it back to the pre-set speed. The governor also prevents excessively high speeds which could damage the engine. Vane and flyweight governors are the more common types of governor used for small petrol engines.

Vane governors, usually found on air-cooled engines, are operated by the airflow from the cooling fins on the engine flywheel. A linkage connects the vane to the throttle butterfly valve, and when the engine runs faster than the pre-set level, the increased airflow causes the vane to actuate the linkage and close the throttle by

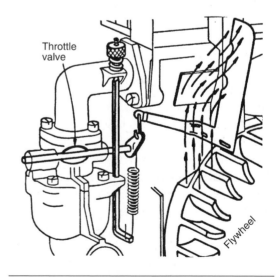

Figure 8.6   A vane-type governor mechanism. (Briggs & Stratton)

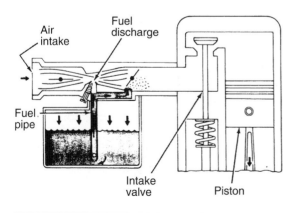

Figure 8.5   A suction lift carburettor. (John Deere)

a small amount. When the engine speed falls away, a spring mechanism on the vane opens the throttle butterfly valve to increase the supply of fuel and air.

A flyweight governor works in a similar way, but it has a pair of hinged weights, usually on the flywheel or the camshaft, with a spring-loaded linkage connecting it to the throttle butterfly valve. The weights swing outwards on their pivots when the engine speed increases above the pre-set level and move inwards when the speed falls away. The linkage opens or closes the throttle valve to vary the quantity of air and fuel mixture supplied to the engine and maintain the selected speed.

The choke is an adjustable valve between the air cleaner and the carburettor. It is used to start a cold engine by restricting the airflow to the engine, and in this way supplies an air/fuel mixture with a higher petrol content. This is known as a rich mixture.

Some engines have an automatic choke, usually consisting of a spring-loaded diaphragm in the pipe between the air cleaner and the engine. When the engine is being started, the spring holds the diaphragm in its closed position, which restricts airflow and gives a richer mixture. When the engine is running, the vacuum created by the venturi in the air tube to the engine compresses the diaphragm spring and disengages the choke.

Air/Fuel Ratio

For normal running, the carburettor is set to mix approximately 1 part of petrol with 15 parts of air by weight. This means that 1 kg of petrol is mixed with 15 kg of air to give an air/fuel ratio of 15:1. When the choke is used, the carburettor supplies a much richer mixture to the engine with an air/fuel ratio of 8:1. It is difficult to understand air/fuel ratio by weight, but in terms of volume, with a fuel ratio of 15:1 about 9,000 litres of air will be mixed with 1 litre of petrol. An engine will not develop full power if it is set to run on a weak mixture

such as 17:1. A rich mixture, e.g. 12:1, wastes fuel and results in a build-up of carbon around the valves and in the combustion space above the piston. Excessive carbon deposits reduce engine efficiency.

Smoke signals from the exhaust give clues about the way an engine is performing. A rich mixture usually causes black smoke from the exhaust pipe when the engine is at its normal running temperature. Black smoke will also appear if the choke is still in use. It can also be caused by a clogged air cleaner which restricts the air supply to the engine.

Don't confuse black smoke with blue smoke, which is a sign of an engine using (burning) oil. This occurs when the engine is worn. Blue smoke may also appear if too much oil is mixed with the petrol.

Care of the Fuel System

It is important to use clean fuel and handle it in clean containers. Many small engine fuel systems have a filter and a sediment bowl, often combined with the fuel tap on the tank, to trap any water and dirt present in the fuel. This needs to be cleaned periodically. The carburettor float chamber will collect dirt after a time. The float chamber should be removed from time to time, taking care not to damage the paper sealing gasket, and cleaned with petrol. Do not clean the float chamber with a fluffy rag. Re-assemble the carburettor, making sure that the gasket is in good condition and correctly fitted.

The carburettor jets are very small, so it is important to use clean fuel to reduce the risk of a blockage. Never attempt to clear a blocked jet with a piece of wire as this may enlarge the diameter of the jet. A blocked jet should be cleared with compressed air.

Petrol will become stale if it is left in the fuel tank for a long period, and an engine that has not run for many months may be difficult to start. The remedy is to drain the tank and refill it with fresh fuel.

When filling a petrol tank, make sure there are no naked flames nearby, and do not smoke. A very small spark can ignite petrol fumes. Make sure there is plenty of ventilation when running an engine inside a building. After servicing, it is better the move the machine into the open air before starting the engine.

THE AIR CLEANER

Almost all small engines have a dry air cleaner, but a few of them which are used in very dusty conditions may have an oil-bath air cleaner (see page 24). Dry air cleaners are usually attached to the carburettor. The filter element, which may be made from paper, felt or fine plastic gauze, is contained in a metal or plastic housing.

Many manufacturers advise servicing at intervals of 25 hours in normal working conditions and more frequently when working in a very dusty atmosphere.

There is considerable variation in the method of servicing small engine air cleaners, and guidance should be taken from the operator's manual. As a general rule, filter elements made of paper may be tapped gently against a flat surface to remove loose dust, but they should not be washed or cleaned with compressed air. Soiled paper filter elements should be replaced

with new ones. Some foam rubber and felt filter elements can be washed with liquid detergent and water and carefully dried.

Small engines may have a two-part air cleaner consisting of a pre-cleaner gauze and a main filter element. The pre-cleaner should be washed with liquid detergent and water and dried with a cloth. After saturating the pre-cleaner in oil, it should be squeezed in a dry cloth to remove surplus oil before using it. Dust can be removed from the main element by tapping it against a flat surface. A typical air cleaner of this type requires the pre-cleaner to be serviced every 25 hours, and the main element at 100-hour intervals.

LUBRICATION

Unlike multi-cylinder engines with a pressurised lubrication system as described in Chapter 2, most small four-stroke engines have a splash feed lubrication. Some more powerful single- and twin-cylinder engines may have pressure lubrication with a pump to circulate the oil around the engine.

Splash Feed Lubrication
The bearings and other moving parts are lubricated by oil splashed over them by a small dipper scoop attached to the big end bearing cap.

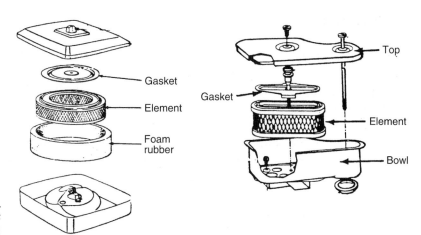

Figure 8.7   Types of small engine dry air cleaner. (Briggs & Stratton)

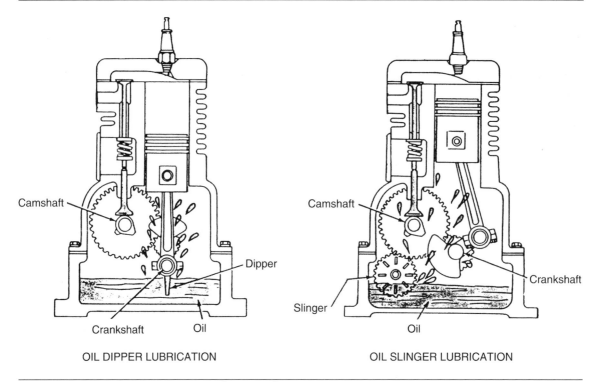

Camshaft — Dipper — Crankshaft — Oil

OIL DIPPER LUBRICATION

Camshaft — Crankshaft — Slinger — Oil

OIL SLINGER LUBRICATION

Figure 8.8   Splash feed lubrication systems. (John Deere)

As the engine rotates, the dipper scoop collects oil from a trough at the bottom of the crankcase. Instead of a dipper scoop, which splashes oil around the engine, some small engines have an oil slinger or a small paddle wheel pump attached to the timing gears.

The piston oil scraper ring (see page 14) collects oil splashed onto the cylinder walls on the piston and passes it through holes in the scraper ring groove to lubricate the gudgeon pin, which holds the piston on to the connecting rod. The oil also runs down the connecting rod and through a small hole to lubricate the big end bearing and then returns to the bottom of the crankcase. The main bearings and timing gears are lubricated by oil which is splashed around inside the crankcase.

It is important to maintain the oil at the correct level in the crankcase to ensure that the splash feed system provides full lubrication to all parts of the engine. This type of lubrication system is not suitable for applications which entail the engine being tipped sideways, as the bearings may be starved of oil.

MAINTENANCE

Regular oil level checks are important and should be made before starting the engine for the first time each day. Some small engines have an oil check system with an indicator light to warn of low oil level in the crankcase.

- The oil must be renewed at regular intervals: 25 hours is a typical change period, but this will vary with different engines. Check with the engine instruction book to find the correct oil change period. Some small engines have an oil filter which should be renewed when changing the oil.

Oil Mist Lubrication

This system is used to lubricate two-stroke engines. As the crankcase must be dry and air-tight, the moving parts are lubricated by oil mixed with the petrol. When the mixture of fuel and air is drawn into the crankcase, an oil mist is created, and this lubricates the moving parts of the engine. The amount of oil mixed with the petrol will depend on the type of engine and the quality of the lubricant. The mixture of petrol and oil can vary from 16 parts of fuel to 1 part of oil up to as much as 60 parts of fuel to 1 part of oil. A typical chain saw engine requires one part of oil to 40 of petrol when high-quality two-stroke oil is used. The mixture for the same engine when using lower quality oil is reduced to 25 parts of petrol to 1 part of oil.

An advantage of an engine with oil mist lubrication, especially those used for chain saws, is that it can be used in any position, even upside down, since all the moving parts will still be lubricated.

Some two-stroke engines have an oil-injection lubrication system. The fuel tank only contains petrol and a second tank holds the oil. An engine-driven pump injects a controlled stream of oil into the mixture of fuel and air as it is drawn into the cylinder.

Check with the instruction book to find the correct mix for any engine you may use. Measure the required quantity of oil and put it into a clean can. Add the correct amount of petrol and shake the contents to ensure thorough mixing. Only mix sufficient oil and fuel for about one week's supply, as stale fuel can make a two-stroke engine difficult to start. Always follow the fuel mixture instructions given in the operator's manual. The addition of extra oil will result in carbon deposits building up inside the engine and lots of blue smoke from the exhaust pipe.

COOLING

All modern small engines are air-cooled. There is no risk of frost damage, but lack of maintenance may result in serious overheating, and in extreme circumstances the engine may seize. Good maintenance is particularly important for two-stroke engines, especially those on chain saws which run at very high speeds.

The cylinder block and cylinder head have cooling fins that conduct the heat away from the engine. The engine flywheel has fan blades which blow air over the cooling fins. A metal cover, or cowling, surrounds the fan blades and fins. This directs the air over the cooling fins, and in this way it removes the surplus engine heat.

It is very important to keep the fins clean to ensure the cooling system is efficient in removing heat from the engine. It will be necessary to remove the cowling at intervals to brush or blow any dirt and other debris from the spaces between the cooling fins. Blocked fins reduce cooling efficiency and the engine will overheat. Oily deposits on the fins, either through a leak

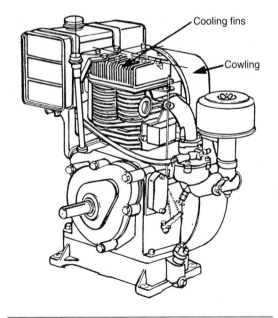

Figure 8.9   The engine cowling has been partly cut away on this air-cooled engine to show the cooling fins around the cylinder. (Briggs & Stratton)

or accidental spillage, should be wiped away as the oil will attract dirt and block the cooling fins. The air inlets in the engine cowling must be kept clear to ensure maximum airflow from the flywheel blades.

IGNITION SYSTEMS

The spark, which ignites the compressed mixture of air and fuel in the cylinder, must be timed to occur just before TDC on the compression stroke of the four-stroke cycle. A magneto or coil ignition system converts low voltage electric current into high voltage current, which is supplied to the sparking plug at the correct point in the compression stroke. Single-cylinder engines have a magneto, and some multi-cylinder engines have coil ignition. A magneto generates its own low voltage current, whereas a coil ignition system needs a battery.

Electricity generated when a flywheel magneto is rotating flows from the coil through the closed contact breaker points to earth. When the cam opens the contact breaker points, the low voltage current can no longer pass through the primary circuit. At this point, a brief pulse of high voltage current of between 10,000 and 15,000 volts is induced (created) in the secondary circuit of the coil. This brief pulse of high voltage current flows through the plug lead to the sparking plug.

The Magneto

The magneto generates its own low voltage electricity and converts it into high voltage current – between 10,000 and 15,000 volts – which is then passed to the sparking plug at the correct time in the four-stroke cycle. The main parts of a flywheel magneto on a small engine are the engine flywheel, contact breaker points, coil and condenser. A switch is usually provided to earth out the current flow to the sparking plug and stop the engine.

Electricity is generated when an electrical conductor is rotated in a magnetic field.

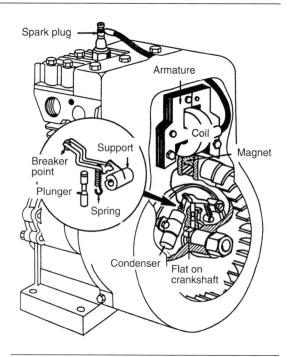

Figure 8.10   A flywheel magneto. Inset shows detail of the moving contact breaker point and its return spring. (Briggs & Stratton)

Alternatively, the magnet can rotate around a static conductor (this principle is explained on page 34). Most single-cylinder engines have a magneto built into the flywheel and driven by the engine crankshaft. Magnets on the inside rim of the flywheel rotate in close proximity to the coil, a static conductor, which is attached to the engine housing inside the flywheel. The coil is a type of transformer that converts the low voltage current, produced by the magneto, into high voltage current, when flow in the low voltage circuit is interrupted by the contact breaker points.

The coil has two separate circuits. The low voltage or primary circuit, which includes the contact breaker points, has a coil with a small number of turns of insulated wire (windings) around a central iron core. The coil in the high voltage or secondary circuit, which feeds current to the sparking plug, has a very large number of windings wrapped around the primary coil. There is a layer of insulation between the primary and secondary windings, which are not connected in any way.

In a standard transformer used to increase or decrease the voltage supplied to an electrical appliance, the output voltage is related to the number of windings on the secondary coil. To illustrate this, a transformer having a primary coil with 200 turns of wire and a secondary coil with 400 turns will give an output voltage twice than of the input voltage. Thus, with an input of 12 volts, the output voltage would be 24. The secondary coil in a magneto will have many thousands of windings to give the very high voltage output needed at the sparking plug.

The contact breaker points are a form of high-speed electric switch in the low voltage or primary circuit. They are opened by a cam on the magneto drive shaft and closed by spring pressure. Many flywheel magnetos have a breakerless system: this means that the low voltage circuit is controlled electronically. The contact breaker points should have a gap of between 0.3 and 0.5 mm when they are fully open. Timing of the spark is achieved by the cam, which is usually an integral part of the crankshaft.

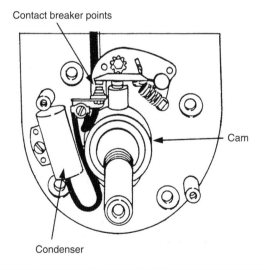

Figure 8.11   The contact breaker points and condenser in a flywheel magneto. (Briggs & Stratton)

When the cam opens the contact breaker points, the sudden interruption of flow in the primary circuit would cause arcing (sparking) across the faces of the points. This is undesirable because the tungsten faces of the points would rapidly burn away. The condenser, which is part of the primary circuit, overcomes this problem. It is a form of electrical shock absorber which reduces arcing when the contact breaker points open. It absorbs the current which is used to boost the strength of the current induced in the secondary winding (coil). The contact breaker points on a flywheel magneto should have a gap of between 0.3 and 0.5 mm when they are fully open. Timing of the spark is achieved by a cam, which is usually an integral part of the crankshaft. Most flywheel magnetos can be fitted to the crankshaft in only one position, as a small key in the shaft has to be aligned with the keyway in the flywheel hub before it is secured in position with a nut.

Electronic Ignition Systems

Contact breaker points suffer from mechanical wear and the contact faces become pitted or burnt with use. This means that frequent servicing and replacement is necessary. At high operating speeds, the contact breaker points also tend to bounce, and this can result in incorrect timing of the spark.

Electronic ignition systems work on the same electrical principles, but the contact breaker points are replaced with an electronic trigger switch and rotor. This unit is easy to replace if it fails, and overcomes the problems of pitting, bounce and wear on contact breaker points.

Sparking Plugs

A sparking plug has a metal body which is screwed into the cylinder head. The central part of the plug is an electrical insulator which is ceramic. A thin metal rod, called an electrode, running through the centre of the ceramic material carries the high voltage current from

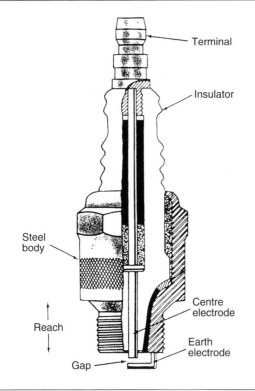

Figure 8.12   Section of a sparking plug. (John Deere)

the magneto to a second, or earth, electrode, attached to the bottom of the sparking plug body. A spark occurs when the current jumps from the centre electrode to earth at the correct point in the two- or four-stroke cycle.

There are many types of sparking plug, made to suit different engines and working temperatures. They are made with three different lengths of thread, or reach, most being 14 mm in diameter. The engine operating temperature also influences the choice of plug. Some engines need cold running plugs with a metal body designed to take the heat away from the electrode. Hot running plugs have a shorter body which keeps the electrode at a high temperature. The sparking plug heat range indicates whether the plug is meant for hot or cold running.

It is important to use the correct sparking plug as specified in the owner's handbook. An engine fitted with a cold running plug will not run efficiently if it needs a hot running plug. The reach must also be correct. For example, a long reach plug used where a short reach plug is required will reduce engine efficiency, especially with small single-cylinder engines. It is important to fit a matching set of plugs in a multi-cylinder engine, as an engine with odd plugs will waste fuel and lose power.

Ignition Troubles

Most starting and running problems are due to fuel or ignition faults. Provided there is plenty of fuel in the tank and the carburettor is clean, look for the following minor faults in the ignition system before seeking help.

1. When there is no spark, it may be due to:
   • dampness in the magneto
   • contact breaker points stuck open or shut
   • very dirty or pitted contact breaker points
   • a broken or loose wire
   • faulty sparking plug
2. When there is a very weak spark, look for:
   • dirty spark plug or incorrect gap setting
   • dirty contact breaker points
   • loose wires or terminals
   • dampness

STARTERS

Some single and twin-cylinder four-stroke engines have an electric starter, but recoil starters are used on two-stroke and many single-cylinder four-stroke power units.

Recoil Starters

Cord-operated recoil starters have a coil spring which is tensioned when the cord is pulled to start the engine, and when the cord handle is released, the spring tension rewinds the cord. When the cord is pulled, two retractable pegs on the recoil starter drum engage with slots in the crankshaft pulley and turn the engine.

The recoil starter cord and spring will give long trouble-free service by following these simple rules:

1. Pull the cord handle slowly until the starter pegs engage, then pull rapidly to avoid kick back on the cord.
2. Do not pull the cord in rapid succession. After each pull, allow the cord to rewind in a controlled way before pulling it again.
3. The cord should not be pulled out to its maximum extent.

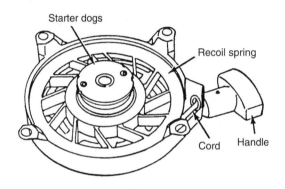

Figure 8.13   A recoil starter. (Briggs & Stratton)

There are few problems with recoil starters. The whole unit can be removed from the engine cowling for cleaning by undoing three or four nuts or studs. A broken cord is not difficult to renew; take the starter off the engine, remove the broken pieces and fit a new cord.

Electric Starters

Some small engines have an electric starter motor powered by a 6- or 12-volt battery. The 6-volt system has a battery that in many cases has to be periodically recharged with a small mains electric charging unit, normally supplied with the engine. Engines with a 12-volt starter have a battery and a generator; this may either be a small alternator (see page 34) or a starter motor with a built-in generator unit. Most small engine starter motors are the inertia type, with a pinion which moves along the starter motor shaft and engages with a ring gear on the engine flywheel. The pinion moves along a coarse thread, called a helix, to mesh with the ring gear. When the engine starts, the pinion is thrown out of mesh with the ring gear.

CHAPTER 9

Ploughs

The plough is a primary cultivation tool which has been used in its various forms for many centuries as the first step in the preparation of a seedbed. The main purposes of ploughing include:

- The complete burial of surface trash, crop residues and farmyard manure.
- Improvement of drainage and air movement through the soil.
- Restoration of soil structure damaged by heavy farm machinery.
- Leaving the land in a condition that permits the production of seedbeds with a minimum number of passes with cultivation machinery.

TYPES OF PLOUGH

Reversible and conventional right-handed ploughs may be mounted or semi-mounted and have a number of bodies which cut and turn furrow slices. Four, five and six furrow reversible ploughs are the most common, but ploughs with up to fourteen furrows are used on some large arable farms.

Reversible Ploughs

Reversible ploughing dates back to the days of the steam engine and the horse. Now in almost universal use on British farms, reversible ploughs with right- and left-handed mouldboards work up and down in the same furrow from one side of the field to the other, leaving a level surface which makes it easier to prepare seedbeds. Very little marking out is necessary

Plate 9.1 Three, four and five furrow mounted reversible ploughs with a hydraulic reversing mechanism are used on many small- and medium-sized farms.

before ploughing can start, and there is little idle running on the headlands.

Reversible ploughs may either be mounted or semi-mounted, and many have a depth wheel or wheels at the back of the plough. Mounted ploughs are lifted clear of the ground on the three-point linkage for transport, while the back of semi-mounted models is carried by a depth/transport wheel when in work, on the headland and on the road. The front of a semi-mounted plough is attached to the tractor hydraulic linkage, but the plough is not lifted completely off the ground. Semi-mounted ploughs have either one or two castor wheels at the rear, and on some models these wheels have a steering mechanism connected to the tractor linkage. Multi-furrow semi-mounted ploughs are lifted

Plate 9.2   Semi-mounted multi-furrow ploughs are lifted from work in two stages. The front is raised on the lower lift arms, and after a slight delay the rear wheel ram lifts the back of the plough.

Plate 9.3   On arrival at the headland, a hydraulic ram on this ten furrow articulated plough lifts the front bodies out of work. A second ram lifts the back section of the plough when the rear bodies reach the headland. This plough has a set of three-furrow presses to consolidate the furrows. (Kverneland)

and lowered in two stages. The front of the plough is raised on the lift arms as it reaches the headland, and after a slight delay the back of the plough is lifted by a hydraulic ram on the rear depth wheels. The front of the plough is lowered before the rear at the start of the next run, keeping the ends of the furrows reasonably level and even for ploughing the headlands. Some tractors have a sequential headland management system which automatically lifts the plough out of work at the headland, reverses the bodies and then returns the plough into work at the start of the next bout.

Articulated Ploughs

Another type of multi-furrow semi-mounted plough, usually with eight to fourteen furrows, has a second pair of wheels about two-thirds of the way along the frame. These wheels, carried on a heavy axle, are self-steering and give improved manoeuvrability when turning at the headland. The rear section articulates vertically from pivot points on the wheel axle to allow the bodies to follow the ground contours and keep all of the furrows at the same depth. When entering work, the front bodies are lowered before those at the rear. Some ploughs are controlled

manually, while others have a sequential electronic headland management system. A typical example lifts the front bodies, followed by the rear bodies, and then operates the reversing mechanism. The system steers the plough while turning on the headland, aligns the plough behind the tractor and lowers the bodies back into work. Some articulated ploughs raise and lower the bodies in three stages – front, middle and rear – with this order reversed when the plough returns into work.

Push–Pull Reversible Ploughs

Used on some large arable farms, these consist of one plough mounted on the rear linkage with a second carried on a front three-point linkage. Various combinations are used, but three or four furrows at the front and up to five on the rear linkage is a typical arrangement for a large tractor. The front plough takes the place of the front weights needed when using a rear-mounted plough, but manoeuvring a push-pull plough in a confined area and matching the work done by the front and rear bodies can be difficult.

Plate 9.4   Careful matching of the work done by the separate front- and rear-mounted ploughs is required when using a push-pull reversible plough. Although no longer popular, this type of plough is still used on some large arable farms. (Gregoire-Besson)

Shallow Reversible Ploughs

Ploughing at a depth of 200–300 mm (8–12 in.) is standard practice on most farms. These depths are desirable for root crops, but not necessarily for cereals. Although rarely used, shallow ploughs with specially shaped mouldboards completely invert the soil, turning furrows 60–180 mm (3–7 in.) deep and 300–375 mm (12–14 in.) wide.

Conventional Ploughs

These have right-handed mouldboards. They may be mounted or semi-mounted and turn between one and eight furrows. Conventional ploughs are occasionally used commercially in some areas of Great Britain, but in most parts of the country they will only be seen at ploughing matches. The soil-working parts are set and adjusted in the same way as reversible ploughs.

Disc Ploughs

Rarely seen in Great Britain, these ploughs have large rotating discs instead of mouldboards. The discs cut and turn furrow slices but do not completely bury surface trash. Disc ploughs are common in countries where this does not matter as the trash is soon scorched by the sun. Both right-handed and reversible disc ploughs are made.

REVERSIBLE PLOUGHING

Reversible ploughing dates back to the days of the steam engine and the horse. Now in almost universal use on British farms, the reversible plough has left- and right-handed bodies, enabling it to work up and down in the same furrow from one side of a field to the other. Reversible ploughs may be mounted or semi-mounted and are heavier and more expensive than right-handed ploughs. However, they have the great advantage of leaving a level surface, which makes seedbed preparation and harvesting easier. Very little marking out is necessary before ploughing can begin, and idle running on the headlands is minimal when compared with a right-handed plough.

Driving the tractor with the furrow-side wheels in the furrow bottom provides the most efficient line of draft between the tractor and plough. It is also easier to steer the tractor, and driving with the front wheel against the furrow wall will keep the front furrow at the correct width. This is less satisfactory when using a tractor with very wide tyres. Although they make better use of tractor power, the tyres may compact part of the last furrow slice turned on the previous run. The use of a furrow widener on the rear body will

Plate 9.5   The furrow widener on the rear body of this plough allows wide tractor tyres to run in the furrow bottom without compacting the last furrow slice turned on the previous pass. (Dowdeswell)

help to overcome this problem by leaving more room for the tractor wheels on the next run.

Driving with all four wheels on unploughed land is another way to accommodate wide tractor tyres. This is also done when ploughing with a crawler tractor on either steel or rubber tracks. The plough can be hitched to the tractor in a way that allows the wheels or tracks to run on unbroken ground and pull the plough in correct alignment with the tractor.

Reversing Mechanisms

At the end of each pass, the plough is rotated on its frame to bring the opposite set of bodies into work. The turnover mechanism is operated by a double-acting ram on the headstock that is supplied with oil by the auxiliary services of the tractor hydraulic system. The ram is controlled by a lever or switch in the tractor cab. Many turnover ram controls have an automatic changeover valve which rotates the plough to the opposite set of bodies when the plough is lifted at the headland.

A manual turnover mechanism will be found on some older mounted reversible ploughs. It is loaded by the weight of the implement pushing forward towards the plough headstock. The reversing mechanism, which is actuated with a hand lever, is re-loaded when the ploughing restarts after the plough has been lowered into work.

Plate 9.6 This nine-furrow on-land plough, which is operated with all four of the tractor wheels running on unploughed land, leaves an uncompacted furrow bottom.

Plate 9.7   The turnover ram on the headstock of a mounted reversible plough.

PLOUGH COMPONENTS

The parts of a plough which engage with the soil – the plough body, disc or knife coulter and the skim coulter – are attached to legs or stalks, which are in turn bolted to the plough beam or frame.

The frog forms the base of the plough body, and the soil-wearing parts are bolted to it.

The mouldboard lifts and turns the furrow slice. It is made in various shapes and sizes with each type producing its own furrow profile and surface finish.

The share cuts the bottom of the furrow slice. When worn, it is important to replace it with a new one. The share is shaped so that it pulls itself into the ground, and as it wears away the plough will require more power to pull it

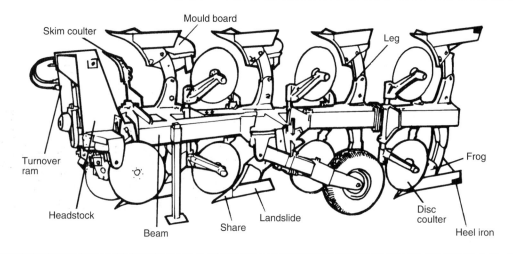

Figure 9.1   The parts of a mounted reversible plough.

through the soil. A plough body with a worn share will not have enough 'suck' (see Figure 9.2) to ensure that it penetrates the ground to the required working depth.

The point and wing form a two-piece share often used on semi-digger and digger bodies. The point makes part of the vertical cut of the furrow wall, and the wing cuts the furrow bottom.

The shin forms the leading edge of the mouldboard on some types of plough body. It takes much of the wear, and when worn is cheaper to replace than a complete mouldboard.

The landside runs alongside the furrow wall and absorbs much of the side thrust of the plough. Most ploughs have a longer landside on the rear body. The landside and share are designed to give 'lead' towards the unploughed land (see Figure 9.2) and maintain the required furrow width. A worn landside and share reduces the amount of lead on the plough body.

A heel iron, which is bolted to the back of the rear landside on some ploughs, helps to carry the back of the plough.

The mouldboard extension, or *tailpiece*, which helps to press down the furrow slice as it leaves

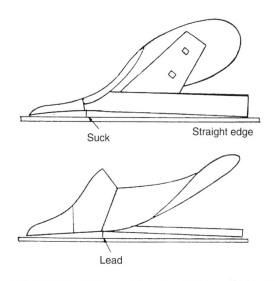

Figure 9.2   A plough body should 'suck' into the ground to maintain its working depth, while the front landsides should 'lead' in the straight line to maintain the front furrow width. New shares are required when plough bodies have no suck, and the front landsides should be replaced when they have no lead.

the mouldboard, is particularly useful when ploughing up grassland.

The disc coulter cuts the side of the furrow wall about to be turned. Some ploughs have serrated or notched edge disc coulters that can be an advantage when ploughing in large amounts of trash or farmyard manure. Adjustment is provided to alter the lateral position and the height of the disc.

A knife coulter serves a similar purpose to a disc coulter and is really a large knife that takes up less space on the plough beam. No longer in common use, knife coulters leave a less defined furrow wall.

Neither disc nor knife coulters are now a standard fitting on all ploughs. However, disc coulters are usually fitted to the rear bodies of multi-furrow reversible ploughs to leave a clean cut furrow wall for the next ploughing run.

Sword shares, with a short vertical knife which cuts part of the wall of the furrow about to be turned, are fitted to some ploughs. Sword shares give more clearance than disc coulters between adjacent plough bodies (see Plate 9.9).

The skim coulter, or *skimmer*, turns a small corner of the furrow slice into the previous furrow bottom before it is turned by the mouldboard. This reduces the likelihood of trash being left exposed between the furrows. Most skim coulters have a small mouldboard and share; others have a one-piece share. There are various sizes of skim coulter, and a large one may be needed when there is a lot of surface vegetation, and for ploughing in heavy dressings of farmyard manure.

Trash boards, fitted above the mouldboards on some plough bodies, improve the burial of surface rubbish and are sometimes used as an alternative to skim coulters.

Safety Devices

Plough bodies used in normal field conditions usually have a shear bolt in each leg to protect them from damage. When the body hits a hidden obstruction, the shear bolt breaks, allowing the body to swing backwards out of work. It is important to use the correct replacement bolt.

Ploughs used in tough or rocky conditions often have an automatic reset mechanism

Plate 9.8   This plough has an auto-reset mechanism on each leg, consisting of a strong leaf spring which allows a body to swing upward if it hits an obstruction. The body automatically returns to the working position after the obstruction is passed. (Kverneland)

on each body which protects it from damage when it hits an obstruction. A heavy leaf spring or a coil spring holds each body in its working position. When a plough body encounters an obstruction, the spring-loaded mechanism allows it to swing back out of work, and after passing the obstruction the spring mechanism returns the body back into work.

Another type of auto-reset mechanism uses an oil and gas accumulator. Shock loads cause the oil to compress the gas, allowing the body to swing backwards out of work; and when the gas expands again it returns to its working position.

TYPES OF PLOUGH BODY

Traditionally, the various types of plough body are classified as general purpose, semi-digger and digger. The surface finish left by the different types of plough body will depend to a great extent on the type of soil being ploughed. Many variations in mouldboard design have appeared over the years, but they still basically conform to the original plough body classification.

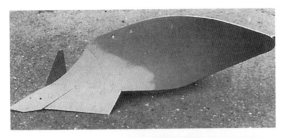

Plate 9.9   Types of plough body. From the top: general purpose with a sword share; semi-digger; digger body for deep ploughing; slatted mouldboard.

General purpose This is a low draft body with a gently curving mouldboard which turns a well-defined furrow three parts wide by two parts deep, e.g. 300 mm wide by 200 mm deep. It is useful for ploughing up grassland and sets up the furrows for weathering by winter frosts, which reduces the time taken to prepare a seedbed for spring sown crops.

Semi-digger This has a shorter mouldboard with a more abrupt curve. It turns an almost square section furrow and leaves a partly broken surface finish. Semi-digger bodies can be used at different depths and speeds, which makes them suitable for most of the general ploughing on the farm.

Digger This body has a short, abruptly curved mouldboard which can turn a furrow deeper that its width and usually leaves a very broken surface finish. Digger ploughs, which have a high horsepower requirement, are mainly used for land to be planted with potatoes and other root crops.

Bar point Rarely used, a bar point body has a special share with an extendable bar, often spring-loaded, which is moved forward as the point wears away. Suitable for some types of plough body, the bar point share is ideal for rocky soil conditions where hidden boulders occur close to the surface. The spring-loaded bar recoils when it strikes an obstruction below the surface and then returns to its working position. Ordinary shares would be damaged in such soil conditions, and although a bar point leaves an untidy furrow bottom, it saves the cost of frequent share replacement.

Continental mouldboards Similar in shape to those used on a semi-digger body, these can be used for a wider range of working depths and ploughing speeds. They are used by British and continental manufacturers and have replaced the more traditional types of mouldboard found on earlier ploughs.

Slatted mouldboards Preferred by some farmers, these consist of a number of curved steel slats bolted to the frog. They tend to break up the soil more than a full mouldboard and give improved soil movement across the mouldboard when working in sticky soils.

Plate 9.10   The slatted mouldboards on this five furrow plough improve soil movement across them, especially when ploughing in sticky soil conditions.

CONTROLS AND ADJUSTMENTS

Many of the adjustments on reversible and conventional ploughs are made in the same way.

Hitching Before attaching a reversible plough to a tractor it is important to check that the wheels are at the correct width for the plough, that the tractor has adequate front weight, there is equal air pressure in the rear tyres and that both hydraulic lift rods are the same length.

Accurate driving to align the hitch points on the plough will make it easier to attach the plough and as most ploughs have automatic linkage there will be no need to leave the tractor cab.

When attaching a mounted plough by hand, connect the lower left-hand lift arm first, then the lower right hand-lift arm and finally the top link. Remember 'left-right-top'. If necessary, use the levelling box on the right-hand lift rod to align the lift arm with the plough cross shaft but make sure both lifts rods are at the same length before starting work. Remove the link arms in the reverse order when detaching the plough. After the lower lift arms and top link have been secured with linch pins, adjust the top link to its mid position for transport and check that the lift rods are still the same length.

It may be necessary to shorten the top link on long multi-furrow ploughs to prevent the back of the implement striking the ground during transport to the field.

As a general rule when ploughing, the check chains should be left slack, allowing some lateral movement, but there are exceptions and the operator's manual should always be consulted when attaching a plough for the first time.

Furrow width Many ploughs have a fixed furrow width, usually 300, 350 or 400 mm (12, 14 16 in.). The plough will be set at one of these widths but can be altered with a toolkit; this is a lengthy task and is rarely done in practice. However there are times when it would be an advantage to alter the furrow width on smaller ploughs. For example, in good conditions the tractor may have enough power to pull wider furrows, thus increasing the work rate. Alterations to the furrow width may also be desirable when there is a need to change ploughing depth, improve burial of trash or to drive at a higher speed to increase the soil crumbling effect of the mouldboards.

Many multi-furrow ploughs have a variable furrow width facility adjusted with either

Plate 9.11   The ram labelled A on this multi-furrow reversible plough is used to vary the furrow width. Ram B brings the plough into correct alignment behind the tractor and ram C is used to reverse the plough bodies.

a hydraulic ram or a mechanical turnbuckle arrangement. Both methods provide a rapid way of altering the furrow width to a new setting by moving the bodies on the plough frame with the aid of a parallel linkage to a new setting. Varying the furrow width can be carried out on the move. Depending on the model of plough the furrow width can be set between 300 to 500 mm (12–20 in). This facility enables the driver to make full use of the tractor power in varying soil conditions by changing the furrow width on the move and leaving the required surface finish. Narrow furrows turned at a faster forward speed usually gives a more broken finish. Ploughing wide furrows where there is a lot of surface trash usually results in fewer blockages and higher work rates can be achieved when conditions are good.

Depth This is adjusted with the hydraulic draft control lever and its stop, or by using the tractor's electronic depth control system. Large multi-furrow reversible ploughs have an adjustable depth wheel on the main frame, while the rear wheels provide additional depth control on semi-mounted models.

Levelling Check that both hydraulic lift rods are set at the same length as instructed in the operator's manual. Separate screw adjustments are provided on the plough headstock for levelling the left-hand and right-hand bodies. With a conventional plough, the levelling box on the right-hand lift rod is used to achieve a level finish.

Pitch On a mounted plough, pitch is adjusted with the top link. Lengthen the top link when the back of the rear landside, or heel iron if fitted, rides clear of the furrow bottom, and shorten the top link if the back of the rear landside makes a deep mark in the furrow bottom.

Insufficient pitch will make plough penetration difficult, and too much pitch will cause the shares to dig in and produce a rough furrow bottom.

Front furrow width Depending on the model and size of the plough, the front furrow width is altered with either a hydraulic ram or two screw adjusters on the headstock. A lever on the cross shaft is used to alter the front furrow width on a conventional right-handed plough. Incorrect front furrow width is often due to faulty setting of the levelling adjustment on the headstock. Most multi-furrow ploughs have a hydraulic ram, which alters the front furrow width by moving the plough sideways on the tractor.

An incorrect front furrow width will result in a very uneven surface finish. It is quite easy to see the consequence of a wide or narrow front furrow width, with a high or low furrow left on each pass of the plough. Incorrect tractor wheel width settings may also be the cause of a wide or narrow front furrow, and the settings for different furrow widths will be found in the plough instruction book. Incorrect alignment of the plough behind the tractor is another likely cause of wide or narrow front furrows. As a general rule, when the wheels are at the recommended track width setting, the top link will run in a straight line behind the tractor. When the plough is in correct alignment with the tractor, the plough will not crab sideways or have a detrimental effect on the steering. Most multi-furrow ploughs have a hydraulic ram which is used to bring the plough into correct alignment with the top link and in a straight line with the tractor.

Disc coulter settings The disc coulter hub should be over the point of the share, and its height will depend on the depth of ploughing – typically with the bottom of the disc about 35 mm above the share. Set the disc higher for deep ploughing. It should always be high enough to prevent the disc bearing housing dragging on the soil and causing undue wear.

The disc should also be set about 12 mm towards the unploughed side of the share. A stepped furrow wall indicates that it is set too far towards the unploughed land. A ragged furrow wall is caused when the disc is set too far towards the ploughed land.

Skim coulter setting This should cut a wedge about 50 mm deep from the corner of the furrow slice about to be turned. When used with a disc coulter, the skim coulter point should be below and behind the disc coulter hub. The disc should not be set too deep as this can make it difficult to achieve the required ploughing depth, especially in hard land.

Plough Maintenance

- Grease all lubrication points daily during the ploughing season.
- Keep nuts and bolts tight at all times.
- Check hydraulic hoses for leaks or damage.
- Replace all worn soil-wearing parts. This is most important with shares or points because when they are worn, the plough will be harder to pull. Some plough bodies have reversible points which can be reversed when worn.

At the end of the season, replace worn parts, coat bright surfaces with a rust-preventative solution, check the condition of hydraulic rams and repair as necessary, and store the plough under cover.

It is important to make sure that when a mounted or semi-mounted plough is raised on the hydraulic linkage for maintenance work, it cannot be lowered accidentally.

PLOUGHING SYSTEMS

A plough which is not properly adjusted will require extra power to pull it through the ground, wasting both time and money.

Always make sure that:

- The plough is correctly attached to the tractor.
- The tractor wheels are at a suitable track setting for the furrow widths of the plough.
- The pitch adjustment is correct.
- The mouldboards are parallel with each other. The mouldboard stays can be adjusted to correct any fault.
- The shares are in good condition.

- The disc coulters and/or the skim coulters, when used, are set deep enough to do their work but not too deep.
- The plough is matched to the tractor, and it does not have more furrows than the tractor can pull with minimal wheelslip.

Reversible Ploughing

Reversible ploughs are used in preference to conventional models almost without exception on British farms. However, an increasing number of farms have switched to minimum-tillage systems for primary cultivations before drilling combinable crops, but ploughing is returning to favour in some areas as a grass weed control method. Ploughing is also a necessary task for growing root crops.

To ensure neat and accurate ploughing of the headlands, the first task is to turn a shallow headland-marking furrow at each end of the field. These furrows serve as a guide to raise and lower the plough at the start and finish of each pass. The distance from the edge of the field to the marking furrow will depend on the size of the plough. A three-furrow plough needs at least 9 m, and one with five furrows requires a headland at least 12 m wide. Larger ploughs require an even wider headland, and using a generous-width headland will reduce wear and tear on the tractor and limit soil compaction when tight turns are required.

It is best to plough the field on its longest dimension, as this will reduce the number of headland turns required. Start at one side of the field and plough one pass of shallow furrows away from the hedge to disturb the vegetation. Then plough in the opposite direction to return the turned furrows back towards the field edge. Continue ploughing across the field and then plough both headlands to complete the job. Some farmers plough a headland-marking furrow all the way round the field, and after ploughing the body of the field, the headlands are ploughed round and round. It is usual for the headlands to be ploughed towards the hedge and away from the hedge in alternate years.

Conventional Ploughing

Ploughing with a right-handed plough is now almost confined to ploughing matches and the occasional small field in livestock farming areas. This type of plough can be used for systematic ploughing. The field is marked out and ploughed in narrow sections or 'lands', and then the headlands are ploughed.

Alternatively, a right-handed plough can be used to plough round and round the field. This can be done by either starting from the headlands and ploughing round and round to the centre of the field or vice versa.

CHAPTER 10

Cultivation Machinery

Power-driven cultivators and harrows, tined implements, disc harrows, rolls and hoes are among the wide selection of tillage equipment available to the farmer. Many of these implements are used after the plough to prepare a seedbed, but minimal cultivation techniques without ploughing are also used, especially in arable farming areas. A seedbed, often up to 8 m wide, can be prepared in a single pass with one of several different combinations of subsoiling tines, rigid and spring tines, discs, levelling harrow and a crumbler roller or packer roller. With a grain drill, coulters and covering tines added to the combination of implements, a crop of cereals, beans or oilseed rape can be sown in a single pass. Cultivation machinery

is also used for stubble breaking, chopping and incorporating crop residues and inter-row cultivations.

SUBSOILERS

Good drainage and free movement of air in the soil are essential for high crop yields. Tractors and heavy farm machinery can be a major cause of soil compaction which restricts the movement of air in the soil and hinders drainage. Heavy rainfall has the same effect.

Subsoilers are used to break compacted ground to improve drainage. Most subsoilers are mounted on the three-point linkage and, depending on the size of the tractor, have between one and five legs with shares at the foot. Normal working depth is between 400 and 500 mm and the distance between the legs varies from 1–2 m. Power requirement is high: a typical

Plate 10.1 A seedbed can be prepared in a single pass with this combination of discs, tines and press roll. Hydraulic rams fold the front and centre sections and lower the wheels for road transport. (Simba)

Plate 10.2 Wings on each leg of this multi-leg subsoiler increase the shattering effect of the implement on the soil. (Sumo)

Plate 10.3   A shear bolt protects the subsoiler leg from damage through overloading by snapping if a certain load/force is exceeded. (Sumo)

three leg subsoiler working at 500 mm will need a tractor of at least 80 kW (110 hp). Trailed subsoilers with up to nine legs and pulled by a tracklayer or a powerful four-wheel drive tractor are used on some farms. Each share has a replaceable point and it is usually possible to fit new shins – the front cutting edge of the leg – when they are worn. A pair of wing shares is often fitted to each share to increase the shattering effect on the subsoil.

Greater shattering of the subsoil is achieved with a vibrating subsoiler share. One type has a power take-off driven vibrator mechanism which agitates the share as it passes through the soil. The agitation has an added shattering effect on the subsoil but does not heave the

ground upward to the same degree as a fixed subsoil share.

Subsoiling is best carried out when the soil is hard and dry after harvest to obtain the maximum shattering effect. It is usual to subsoil at right angles to the intended direction of ploughing.

MOLE PLOUGHS

Heavy land needs periodic draining to reduce the soil water content to a satisfactory level for efficient plant growth. Heavy soils usually have a system of permanent drains with perforated plastic pipes which discharge water into a ditch. Some older drainage systems have clay pipe main drains. Mole ploughs are used to form small tunnels (mole drains) in the soil at an angle to the main plastic or pipe drains. Mole drains, made at depths of between 35 and 55 cm (15–23 in.), collect ground water, which runs into the pipe drainage system and is discharged into a ditch.

Mole ploughs are usually trailed and pulled by a crawler tractor, but lighter models for use on the three-point linkage of a powerful four-wheel drive tractor are also made. A mole plough has a very strong frame that slides along the ground when the machine is in work. A heavy leg, similar to a subsoiler leg, is attached to the frame, and a circular section share with a larger diameter expander on a flexible link is bolted to the leg. The bullet-shaped share makes a tunnel of about 75 mm diameter and the expander presses the soil outwards to form a long-lasting drainage channel.

CHISEL PLOUGHS

The chisel plough, a heavy-duty form of cultivator, is sometimes used as an alternative to a mouldboard plough, especially where there is little surface trash to be buried.

Chisel ploughs with working widths of 2–5 m are mounted on the tractor three-point linkage and working depth is controlled hydraulically.

Plate 10.4   Some farmers use a chisel plough, often in a combination with discs and a press, in place of a mouldboard plough for seedbed creation.

Wider models usually have a wheel at each side to control the working depth of the heavy-duty rigid, spring-loaded or flexible tines. Each tine has a replaceable share, and on most chisel ploughs the share is protected by an overload shear bolt. If the share hits an underground obstruction, the shear bolt will break, allowing the share to swing back and avoid excessive damage. The tines are arranged in two or three staggered rows and the number will depend on the power available, type of work and soil condition.

CULTIVATORS

Tined cultivators have been used to prepare seedbeds since the days of the horse and traction engines. Power-driven harrows and cultivators are used on some arable farms to reduce the number of passes required to create a seedbed. Tined implements and power-driven harrows can also be used in combination with a grain drill to prepare the seedbed and drill in a single pass.

Tined Cultivators

A cultivator has a tractor-mounted frame with a number of tines which break and stir the soil. Working depth is controlled hydraulically, but very wide mounted cultivators usually have depth control wheels. Like most cultivation equipment, a wide range of working widths from 2–8 m or more is available to suit all sizes of tractor.

Cultivators are used for seedbed preparation, especially after ploughing, stubble breaking, general weed control and cultivating between potatoes and other row crops to kill weeds.

Types of Tine

Rigid tines Used for heavier work. The tines are staggered across the frame in two or three rows to allow free passage of soil and reduce blockages.

Spring-loaded tines These are rigid tines held in their working position by heavy springs which allow the tines to lift when they hit a large clod or other obstruction. The springs return the tines to their working position; this helps to shatter the clods.

Spring tines These are flexible, square or flat sectioned, sometimes with a coil spring which vibrates the tine as it passes through the soil. The vibrating movement of the tines will give fast seedbed preparation in most soils.

Plate 10.5   The vibrating action of the spring tines on this cultivator have a shattering effect on the soil. (Gregoire-Besson)

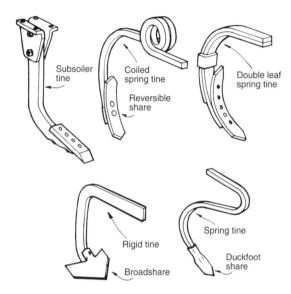

Figure 10.1   Types of tine and share.

Plate 10.7   Hydraulic rams are used to lift the rear trans-port wheels on this combination cultivator and hold them in the raised position when in work. (Vaderstad)

Types of Share

The reversible share is used for seedbed and general work. Duckfoot and broadshares are mainly used for stubble cleaning and breaking. Shares must be replaced when worn – reversible shares have a point at each end and can be turned when the first point is worn away.

Combination Cultivators

Seedbeds can be prepared in a single pass with a mounted or trailed combination cultivator with a mix of tines, discs, presses and levelling boards usually mounted on a heavy wheeled frame. Some farmers use a combination cultivator to prepare a seedbed on previously ploughed land. Others use these implements, often combined with a grain drill, to cultivate

and drill a crop directly into stubble, harvest trash or other unbroken ground.

Minimal tillage techniques, which save both time and money, have become widely used in recent years, especially in arable farming areas. The arrangement of tines, discs, presses and levelling boards varies with different types of combination cultivator, which are made in working widths from 3–10 m or more and require a tractor in the 112–225 kW (150–300 hp) power range. Some designs are better suited to heavy land, while others work best in light soils.

One type of combination cultivator, made in widths between 3 and 7 m, has open cage rollers at the front and rear. A spring-loaded, adjustable levelling board behind the front roller crushes large clods and levels the surface. Rows of spring tines follow the levelling board and the rear open roller gives more consolidation and leaves a level finish. The wider models have hydraulically folded wing sections and a hydraulic ram lowers a pair of transport wheels.

POWER-DRIVEN HARROWS AND CULTIVATORS

Power take-off driven cultivation implements, including power harrows and rotary cultivators, are used to prepare seedbeds on some farms, especially where root crops are being planted and a fine clod-free seedbed is required. Power harrows can also be used in combination with a grain drill to cultivate and drill ploughed land in a single pass.

Plate 10.6   With its tines, levelling discs and packer rollers, this combination cultivator can prepare a seedbed in a single pass. (Vaderstad)

Power-driven cultivation machinery is more expensive to operate than non-powered because of the moving parts and requires regular maintenance to ensure a long and trouble-free working life. An important advantage of power harrows is their ability to prepare a seedbed on ploughed land without bringing unweathered soil to the surface and to leave a relatively clod-free finish. However, power requirement, purchase price and maintenance costs compared with a tined implement must be taken into account.

Power Harrows

There are two types of power harrow with rotary or reciprocating tines. Most farmers use a rotary power harrow, but a few reciprocating tine machines are still in use.

Rotary power harrows have vertical rotating tines attached to a series of gear-driven rotor heads spaced across the full width of the machine. The tines are power take-off driven through a gearbox and a system of gears. Adjacent tine rotors turn in opposite directions and each rotor is timed with its neighbours to prevent them hitting each other. Although power harrows are usually rear-mounted on the three-point linkage, some are attached to the front hydraulic linkage and driven by a front power take-off shaft. This leaves the rear linkage free for a drill, fertiliser broadcaster or a sprayer.

Plate 10.9   The tines, discs and rear press on this cultivator can be used to produce a seedbed from stubble. (Kuhn)

Plate 10.10   Hydraulic rams are used to fold this 4.5 m wide power harrow to within 3 m for transport. (Kuhn)

Plate 10.8   This heavy-duty one-pass cultivator has front and rear gangs of disc harrows, large retractable tines and a rear packer roller. (Teagle)

Plate 10.11   The power take-off shaft extension on this 4 m power harrow, with a rear packer roller, can be used to operate a grain drill mounted on the back of the implement. (Teagle)

Tine rotor speed, which can be varied on most power harrows, ranges between approximately 125–500 rpm. The speed is selected either with a lever on the gearbox or by using different pairs of gears in the gearbox. A rotor speed of 250–500 rpm is recommended for most work. With the exception of some narrow power harrows, most of them are driven with the power take-off in the 1,000 rpm setting. Power requirement is high: a 3 m power harrow requires at least 56 kW (75 hp) at the power take-off, and a 6 m model needs up to 150 kW (200 hp). A slip clutch or shear pin mechanism protects the drive from the power shaft to the main gearbox.

Working widths vary from 1.5–6 m; the smaller lightweight machines are mainly used in horticulture. Power harrows over 4 m wide have a more robust frame and gear drive. Very wide models are either hydraulically folded or end-towed for transport. A 6 m harrow usually consists of two 3 m machines hinged on a central frame with a separate drive to each section. Adjustable soil deflector plates on both

sides of the power harrow frame leave level joins between each pass and on most machines a floating clod comb or backboard follows the tines. Both improve clod breaking performance and a backboard also holds up the soil to give the tines a greater effect on the soil.

Most power harrows have a rear crumbler roller, an open coil roller or a packer roller, which consolidates the soil and regulates the working depth. A crumbler roller works best in good soil conditions. The open coil roller has a greater consolidating effect, and the packer roller compacts the soil in a similar way to a Cambridge roll.

A drill or planter can be attached to an optional three-point linkage frame at the back of a power harrow and driven by an extension from the power shaft. The power harrow and drill are both lifted out of work while turning at the headland. This combination allows a crop to be planted in one pass after the field has been ploughed.

Tilth Adjustment

The fineness of the tilth produced by a rotary power harrow depends on tine speed and tractor forward speed. A fine tilth is obtained with a low forward speed and a high rotor speed. The height of the clod comb or the levelling board also affects the tilth. It should be high enough to let broken soil through but retain the clods for further treatment. Clods will be buried just below the surface if the board is too deep.

Power Harrow Maintenance

Frequent lubrication is important and gearbox oil levels need checking regularly. Make sure all nuts and bolts are tight, paying special attention to the tightness of the tine retaining bolts when used. Some power harrows have quick release clips to secure the tines on the rotor. The tines should be replaced when worn; if they are allowed to become too short, the tine holder may also wear.

Plate 10.12   A power harrow tine and gear drive assembly.

Plate 10.13   Types of rear roller used with power harrows. Top: crumbler roll. Centre: open coil roller with star scrapers. Bottom: packer roller with scrapers.

Plate 10.14   The drill element of this power harrow/drill combination has a mechanical grain feed mechanism. (John Deere/Lemken)

ROTARY CULTIVATORS

Although the power harrow is more popular, the rotary cultivator is often used on farms where potatoes and vegetables are grown. Rotary cultivators, made in working widths from 1 m for use on horticultural holdings to about 5 m for tractors up to 185 kW (250 hp), have a series of power take-off driven 'L' shaped blades bolted in a spiral pattern around a horizontal rotor shaft. Some early rotary cultivators were trailed but they are now mounted on the three-point linkage. Small rotary cultivators are driven from a 540 rpm power take-off, but wider models have a high power requirement and the 1,000 rpm power shaft is normally used.

Drive from the power take-off is through a gearbox and chain drive to the rotor. An overload slip clutch is built into the main drive, and using a different pair of gears in the gearbox (see Figure 10.2) varies the rotor speed. Most rotary cultivators have a gear lever for selecting any one of six or more rotor speeds. The blades are arranged in a scroll pattern around the rotor to maintain a constant load on the tractor power shaft. The blades throw the soil against a hinged flap, and this helps to break up the clods. Heavy rigid tines can be attached to a frame in front of the cultivator to break up the subsoil, and some models also have a rear crumbler roll to consolidate the soil.

Plate 10.15   A rotary cultivator produces a fine tilth when the rear flap is in its lowest position to retain the soil for as long as possible. (Kuhn)

Plate 10.16 The cultivating rotor on this mounted rotary cultivator is power take-off driven through a bevel gearbox and chain and sprocket drive to the rotor shaft. The rear packer roller consolidates the freshly-cultivated soil. (Kuhn)

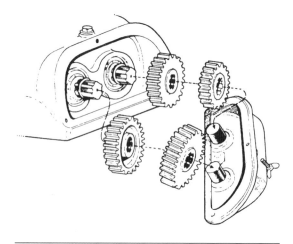

Figure 10.2 One type of gearbox used to change rotor speed on a rotary cultivator.

Adjustments

Tilth Three adjustments can be made to vary the type of tilth produced:

1. The rotor speed can be changed by using a different gear ratio in the cultivator gearbox. A fast rotor speed gives a fine tilth.
2. Changing the forward speed through the tractor gearbox with the power take-off speed, depending on model of rotary cultivator, running at 540 or 1,000 rpm.

3. The hinged flap at the back of the machine is lowered for a fine tilth. It will be very coarse when the flap is used in the raised position and the soil is retained in the machine for less time.

Working depth This is adjusted with a depth wheel; wide rotary cultivators have two depth wheels. An external hydraulic ram is used to vary the working depth of trailed rotary cultivators. A depth-limiting skid is sometimes fitted to prevent the blades going in too deep if the depth wheel(s) drop into a low spot.

Uses of a Rotary Cultivator

- Stubble cleaning.
- Preparing seedbeds, especially for potatoes. A spiked rotor is sometimes used for this work.
- Chopping weeds and crop residues.
- Land reclamation and general tidying around field headlands.

Ground Driven Rotary Cultivators

The tines on this type of rotary cultivator are driven by ground contact. A twin-rotor machine, with the rear rotor chain driven from the front rotor, is still used on some farms for seedbed work and stubble cleaning. Working depth is controlled by the rear crumbler roller, which can be replaced with a packer roller when greater consolidation is required.

DISC HARROWS

Disc harrows cut and consolidate the soil, chop old crop residues and are particularly useful for seedbed preparation on freshly ploughed grassland. Disc harrows may be mounted or trailed and are sometimes used as one element of a combination cultivator. Single gang mounted and semi-mounted disc harrows have one left-handed and one right-handed gang of saucer-shaped discs running side by side. Tandem disc harrows have four gangs of discs with two front gangs and two at

Plate 10.17   Stubble breaking, seedbed work and chopping crop residues are some of the many uses for a set of disc harrows. (Kverneland)

Plate 10.18   The discs and rear crumbler roller on this combination cultivator are folded by hydraulic rams for road transport. (Gregoire-Besson)

the rear. Mounted and semi-mounted disc harrows have working widths of 2.5–6 m. Wide models are folded either hydraulically or manually down to about 3 m for transport. Trailed disc harrows with working widths of up to 10 m or more have large diameter pneumatic tyred transport wheels raised and lowered with hydraulic rams. They are usually three gangs wide, and hydraulic rams are used to fold the wing sections to a vertical position for transport.

Disc diameter varies from 300–750 mm or thereabouts, and each gang of discs is carried

in its frame on ball or taper roller bearings. Some disc harrows have plain discs, others have scalloped discs and some disc harrows have scalloped discs on the front gang and plain discs at the rear. Plain discs have a better cutting and consolidating effect; scallop-edged discs give better penetration in hard ground.

Adjustments

Disc angle The working angle of the saucer-shaped discs determines their cutting effect on the soil. The disc angle is usually altered with a hydraulic ram. Some harrows have a ratchet-operated turnbuckle to adjust disc angle, and older models have a hand lever for this purpose. When set at the widest angle to the direction of travel, the movement of soil will be at its maximum. When running almost straight, the discs have a greater consolidating effect and less movement of the soil.

Working depth This is controlled by the tractor hydraulics and/or ram-operated depth wheels. Some disc harrows have a depth-sensing mechanism connected to the hydraulic ram which maintains the discs at a pre-set working depth. Disc angle will also affect the working depth, and extra weight may have to be added to improve disc penetration when working on hard ground.

Adjustable scrapers Some disc harrows have adjustable disc scrapers which should be set close enough to the discs to keep them clean.

Uses of a Disc Harrow

- Seedbed preparation, especially after ploughing grassland where a tined implement is likely to pull buried turf to the surface.
- Cutting up crop residues.
- Stubble breaking.

Maintenance

Regular greasing and checking the tightness of nuts and bolts are the main servicing needs of disc harrows. The gangs of discs must also be tight on their axles.

SPIKE TOOTH AND CHAIN HARROWS

Spike Tooth Harrows

These consist of a simple frame with tines bolted to it where the frame members cross. There is a variety of harrow sizes and weights, from light harrows for final seedbed preparation to heavy harrows used in some areas for breaking down ploughed land. Traditionally, spike tooth harrows were used in sets of four or more attached to a harrow pole or bar with a wheel at each end and often towed behind a cultivator or a roll. Some farmers use four or five spike tooth harrows suspended by chains from a three-point linkage mounted harrow frame which is folded for transport. It is important to keep harrow tines tight in the frame. Failure to do this will result in worn tine holes and loose tines.

Chain Harrows

Chain harrow links may be spiked or plain. Plain linked chain harrows are sometimes used to spread droppings or molehills on pasture or to cover broadcast seed. Grassland can be aerated with a spike link chain harrow.

ROLLS

Cambridge rolls and flat rolls are used to break clods, consolidate cultivated soil and leave a level surface for drilling. A furrow press, sometimes pulled behind a plough, is similar to a Cambridge roll with the rings spaced further apart.

Cambridge rolls have a number of cast iron rings, 500–700 mm in diameter, on a heavy axle, which leave a corrugated surface finish. Flat rolls are less common and have a number of cast iron rings on an axle with a flat rim. An alternative type has one or more lightweight steel cylinders on an axle or a much heavier cylinder which can be filled with water to make a heavy flat roll for use on grassland.

Trailed three-gang rolls – one wide and two narrow units – with working widths of up to 8 m are used on some farms, but are largely

(A)

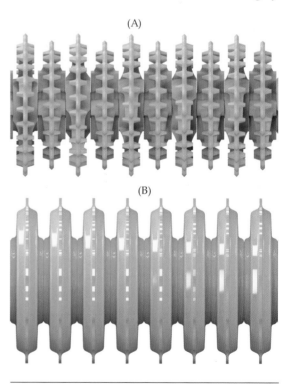

(B)

Plate 10.20   Many Cambridge rolls have ridge-profiled cast iron rings, but others may have heavy Crosskill rings (A), named after their inventor, which break clods more effectively, or light hollow steel rings (B) for work on lighter soils. (Vaderstad)

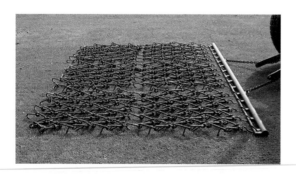

Plate 10.19   Spike link chain harrows are used for aerating grassland. (William Herbert)

Plate 10.21   For road transport, hydraulic rams are used both for folding the outer sections of this set of rolls and for lowering the rear transport wheels. (Vaderstad)

Plate 10.22   Folded for transport, this three-section roll has castellated rings designed for breaking clods. (Twose)

outdated due to the problem of taking them on road and the development of hydraulically folding rolls. Trailed rolls are hitched in-line for transport and, to prevent damage, should only be towed at low speed on the public highway.

Much wider three, five or seven gang rolls with working widths of up to 18 m or more and pneumatic tyred transport wheels are now used on many farms. Hydraulic rams are used to fold the outer gangs, either in a vertical position or inwards alongside the frame for transport.

Maintenance

The axle bearings need regular lubrication when in use. The roll sections should be kept tight on the axle and hydraulic hose couplings should be checked for wear.

FURROW PRESSES

Retention of soil moisture, especially when ploughing in dry conditions, is important for the successful establishment of autumn-sown crops. Rolling as soon as possible after ploughing is one solution to this problem. A furrow press, similar to a narrow version of a Cambridge roll, but with a single or double row of rings which are more widely spaced, can be pulled behind a plough to consolidate the soil ploughed on the previous pass.

The furrow press may either be pulled from a special linkage attached to the plough beam or mounted to the hydraulic linkage at the front of a tractor. Furrow presses with working widths of 1–4 m are suitable for three to seven furrow ploughs. They can also be used on larger ploughs by attaching two or three separate units at intervals along the length of the plough. Trailed furrow presses either have a three-point linkage hitch for transport purposes or can be lifted above the plough with a hydraulic ram and carried in that position.

In operation, a rear-attached furrow press is automatically detached when the plough is lifted out of work at the headland. The press has a double pick-up arm which provides a method of connecting it to a reversible plough in both directions of travel. After turning on the headland, the plough is lowered into work, and a towing hook on the plough engages with the furrow press pick-up arm.

Furrow presses can also be used on a tractor front hydraulic linkage or suspended from a heavy-duty arm which pivots on a frame attached to the front of the tractor. The arm-mounted press is moved from the left- or right-hand side of the tractor with a hydraulic ram

Plate 10.23 This furrow press, trailed by the plough to consolidate the previous pass of freshly ploughed land, has two sets of spring tines to stir the soil and help to break down the furrows. (Kverneland)

Plate 10.24 Ploughing with a reversible plough and furrow press speeds up seed bed preparation by combining two operations in one.

and is held centrally in front of the tractor for transport. An alternative type of front-mounted press, carried on the three-point linkage, can be used to consolidate the land between the front wheels when a drill is being used on the rear linkage.

TOOLBARS

Toolbars can be front-, mid- or rear-mounted. Various implements including hoes, cultivating tines, ridging bodies, etc can be fitted to suit various row crop widths. The tractor wheels must also be set to suit these row widths.

Inter-row hoes may be attached to a front-, mid- or rear-mounted toolbar. Front toolbars are very sensitive to slight movements of the steering wheel, but mid-mounted toolbars are less sensitive to steering changes and provide the driver with reasonable visibility of the rows being hoed. When hoeing with a rear-mounted toolbar, a second operator can be seated on the hoe, to steer it along the rows.

Side hoe or 'L' shaped hoe blades are used to cut weeds close to the rows, and 'A' hoe blades cut through the weeds growing between them. Some inter-row hoes have plant protection shields on the 'L' hoes to prevent them throwing soil over the young plants.

Traditionally, front-, mid- or rear-mounted hoes were used to control weeds in sugar beet and other row crops. However, with the widespread use of chemical weed control for most farm crops, tractor-mounted hoes are rarely

Plate 10.25 Inter-row cultivating a crop of organic wheat grown in widely spaced rows. The toolbar is equipped with an electronic guidance system. (Garford)

used. The increased popularity of organic crop production has resulted in the introduction of a new generation of tractor-mounted toolbars with hoe blades or cultivator tines and working widths of up to 9 m. Some of these toolbars, wide enough to hoe eighteen rows of sugar beet, have independently mounted hoe units often with zero pressure tyres and a seat for an

Plate 10.26   The electronic vision guidance system with in-cab monitoring screen on this inter-row hoe follows the plant rows to ensure accurate weeding. (Garford)

Plate 10.27   A rear-mounted inter-row hoe with shields to protect the young plants. (Garford)

operator to steer the hoe along the rows. Others have an electronic guidance system with a video camera, an automatic side shift mechanism to keep the toolbar on the rows and a monitoring screen in the tractor cab.

WEEDERS AND STRAW HARROWS

Tractor-mounted weeders with spring tines about 6 mm in diameter, which pull out small weeds in growing crops, are used on some farms. Closely spaced tines are clamped to a three-point linkage mounted frame up to 8 m wide, which folds for transport. The position of the tines can be adjusted to suit different row widths, and when used in potatoes, the tines can be set to follow the contours of the potato ridges.

Straw harrows are similar to weeders, but have much stronger tines and can be used at high speed on old stubbles to loosen and level the surface, and in so doing stimulate weed growth in the stubble and destroy slug habitats. The weed seedlings are either sprayed off or destroyed with a second pass using a straw harrow, leaving the field ready for the drill.

Plate 10.28   The tines on this straw harrow, angled with a hydraulic ram, create a shallow tilth to destroy slugs and encourage weed seed germination on stubble before spraying off. (Claydon)

CHAPTER 11

Drills

GRAIN DRILLS

Mounted and trailed drills for grain and other combinable crops can be broadly categorised into three main groups. The first, the grain drill, is used on prepared seedbeds, while the second, often called a minimum-tillage (min-till) drill, is similar, but is combined with a powered or non-powered cultivation implement to break down partly prepared land ahead of the coulters which place the seed. The third is the direct drill, of which there are two types, both used on unbroken ground. One type, based on heavy-duty discs or tines, cuts slots or furrows in the soil; the seed is then dropped into the slots and the soil is then firmed over. The strip-till drill, a more recent development, has cultivation units ahead of the coulters which loosen the soil to improve drainage and encourage root growth. The seed is sown in narrow bands and an unbroken strip of ground is left between each row to support later machinery passes.

All types of drill may have either a mechanical or pneumatic seed metering and delivery mechanism; the latter is more common on drills with sowing widths wider than 3 m which have folding wing sections. Drills with working widths in the 2.5 to 6 m bracket are used on many farms, but trailed models up to 12 m wide are used in some arable farming areas. Unless the drill can be towed end-ways, drills wider than 3 m must be hydraulically folded for transport.

Most mechanical-force feed grain drills, which may be mounted on the tractor three-point

Plate 11.1 The sowing rate is calibrated electronically on this cultivator drill, which features a disc harrow to prepare the soil. Seed feed to the coulters is pneumatic, while an open coil press then firms the soil. (Teagle)

linkage or form part of a mounted cultivator/drill combination, have small pneumatic tyred drive wheels. Older trailed force feed drills of this type usually have large diameter wheels, both to carry the drill and drive the seed metering mechanism. The metering system or feed mechanism in the bottom of the hopper or seed box supplies seed in a controlled way to flexible seed tubes. These direct the seed to disc or Suffolk type coulters, which make shallow furrows in the soil. The seed tubes are made from plastic, but some older drills with mechanical-feed mechanisms have rubber seed tubes. Most drills have a row of closely spaced spring tines behind the coulters and across the full width of the drill to cover the seed.

Although the coulters on trailed drills are usually raised and lowered into and out of work

Plate 11.2   A power harrow and mechanical-feed grain drill combination sowing a cereal crop into unbroken stubble. (Kuhn)

Plate 11.3   A cultivator and drill combination, with spring tines followed by a pneumatic feed grain drill. (Gregoire-Besson)

hydraulically, a mechanical lift clutch does this work on some older drills. The action of raising and lowering the coulters also controls the drive to the force feed seed mechanisms. Seed flow on fully mounted drills is disengaged when the coulters and drive wheels are raised from the ground.

CULTIVATOR AND DRILL COMBINATIONS

Mounted and trailed cultivator/drill combinations are used on many farms, either on previously ploughed land or sometimes directly on stubble. The ground will be cultivated with a combination of tines, discs and packer rollers or a power harrow and packer roller before the seed is drilled. Mounted cultivator/drills consist of a drill unit mounted above either a combination of tines and discs or a power harrow.

The grain feed mechanism in the bottom of the hopper is sometimes chain driven from a land wheel, but on later drills it may be driven by a hydraulic motor. The fan on a pneumatic or air-feed drill drive is power take-off driven. Most mounted cultivator/drill combinations have a drilling width of 3, 4 or 6 m and a grain hopper capacity of 1,000 to 3,000 litres.

Power harrows have a high power requirement and, where soils are suitable, many farmers prefer to use a more economical tined cultivator or disc harrow/drill combination, or carry out cultivations separately.

Trailed and semi-mounted cultivator/drill combinations with various implement arrangements and a drill with an air or mechanical-feed mechanism mounted on a wheeled chassis have higher work rates than mounted models. These drills, with working widths of 4 to 12 m and hopper capacities from 2,000 to 6,000 litres, are used on unploughed or cultivated land and on cloddy or trashy surfaces. An example of this type of drill (see Figure 11.1) is carried on the

Plate 11.4   Seedbed preparation and drilling are completed in one operation with this power harrow, press and pneumatic feed drill with a front-mounted grain hopper. (Kuhn)

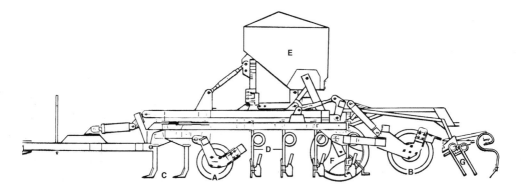

Figure 11.1   A 3 m version of this high output combination cultivator drill requires a 75 kW (100 hp) tractor. The drill has a mechanical seed metering mechanism and the seed is delivered pneumatically to the coulters. (Simba Great Plains)

packer roller (A) at the front, and zero pressure rubber-tyred rollers (B) at the rear. Two rows of cultivator tines (C) are in front of the packer roller, and four staggered rows of coulters (D) follow it. A land wheel (F) drives the mechanisms in the hopper (E), and plastic tubes carry the seed to the coulters. The seed is covered by a row of independently mounted tines (G). The hopper is attached to the drill chassis by a closed hydraulic circuit suspension system so that its weight is equally distributed across the full width of the drill. This gives an even sowing depth to all of the coulters, no matter how much grain there is in the hopper.

STRIP-TILL DRILLS

This type of one-pass cultivator/drill combination cultivates strips of land across the drill's sowing width, leaving unbroken ground between each strip. Strip-till drills have a number of individual cultivating and pneumatic seeding units across their width. A constant pressure hydraulic system is used to hold the units in full contact with the ground. A strip-till drill may have either discs or tines to loosen the strips of unbroken ground. Some strip-till drills also have a row of pneumatic tyres or a crumbler roll used to consolidate the soil before the coulters place the seed from the pneumatic feed mechanism in rows or bands

Plate 11.5   Strip-till drills, used on unbroken ground, cultivate only the strips of soil where seed is to be planted. The undisturbed soil strips in between reduce weed germination and support subsequent field traffic.

Plate 11.6   A crop of oilseed rape drilled with a strip-till drill. (Vaderstad)

up to 150 mm wide. Covering discs or tines then cover the seed with a thin layer of soil, or alternatively they may be firmed in with rubber-tyred press wheels. Strip-till drills have working widths of between 3 and 12 m, with 9 to 36 seeder units respectively. As this type of drill is used on undisturbed ground and it only cultivates the strip of land where seed is planted, it may require less tractor horsepower than a cultivator/drill which cultivates across its full width. Optional equipment for some strip-till drills includes liquid or solid fertiliser placement units or granular applicators.

DIRECT DRILLS

Farmers with easy working soil may use a cultivator or strip-till drill to sow combinable crops into unbroken stubble, while others use a direct drill, which has heavy-duty components to do this work regardless of soil type.

Direct drills have very robust spring-loaded coulters, based on tines or single- or double-angled discs, to cut slots in the soil. Coulters in their shadow place the seed. The coulters run close up against single-angled discs. Double-disc coulters are angled towards each other with the coulter running between

Plate 11.8   Mounted on the three-point linkage frame on a rotary cultivator, this grain drill has land wheel drive to the fluted roller grain feed mechanisms. (Kuhn)

them. Press wheels or tines running behind the coulters cover the seeds.

To reduce establishment costs, oilseed rape is sometimes sown with a broadcaster, which is another form of direct drilling. A seed hopper with either a mechanical or pneumatic seeding mechanism mounted above a combination cultivator, or subsoiler, broadcasts seed in bands or across the full width of the implement. Weed control can be a problem with direct drilling and it is usual to spray off surface vegetation a few days before the field is drilled to eliminate competition for the emerging crop.

Combine Drills

Practised widely in the past, the concept of drilling combinable crops along with fertiliser has recently started to attract interest again as a method of improving crop establishment. Most mechanical force-feed combine drills have a fluted roll or peg roll feed mechanism for the fertiliser which is similar to that for the seed, but the hopper is much larger with separate sections for the grain and the fertiliser.

Combine drills with a pneumatic feed have separate metering units for the grain and fertiliser which are carried in separate compartments in

Plate 11.7   Cereal crops can be drilled into unbroken ground with this 6 m pneumatic feed direct drill with a large capacity grain hopper. (Accord)

the hopper. Alternatively, seed can be carried in a front hopper, from where it is blown to a cyclone which removes the air before the seed is passed to the metering mechanism on the drill. A second type of feed mechanism meters the fertiliser from the rear hopper at the required rate. The air stream then conveys the seed and fertiliser to the coulters.

MECHANICAL AND PNEUMATIC FEED MECHANISMS

Fluted roller feed is used for most mechanical external force feed drills. Pneumatic feed drills use a stream of air to carry the seed from the feed mechanism to the coulters. The feed mechanism has a shaft under the seed hopper with a number of revolving fluted rolls, one for each coulter. The flutes carry grain from the hopper to the flexible plastic seed tubes at a pre-set rate. Many fluted roller drills are only suitable for cereal crops, but some designs have an adjustable shutter close to the rollers. The clearance between the roller and the shutter can be varied to suit different sizes of seed.

Seed rate is altered by:

1. Changing the length of the fluted roller in contact with the seed. A lever is used to adjust the amount of fluting carrying the seed from the hopper. A graduated scale provides a guide for setting the seed rate.
2. Changing the speed of the fluted rollers via a gearbox. The highest seed rate is achieved with the highest roller speed and the full length of the fluted roller in contact with the seed.

Studded roller feed is another type of external force feed mechanism. Studded or pegged feed rollers, also on a shaft in the bottom of the hopper, carry seed to the seed tubes as they turn on their shaft. This type of feed mechanism is suitable for most types of seed.

Seed rate is adjusted by:

1. Changing the speed of the studded rollers.
2. The position of an adjustable shutter which controls the flow of seed from the hopper to the feed mechanism.

Plate 11.9   This 4 m wide power harrow and drill combination, with two staggered rows of coulters, has a fluted roller feed mechanism driven by land wheels at each side of the drill. (Kuhn)

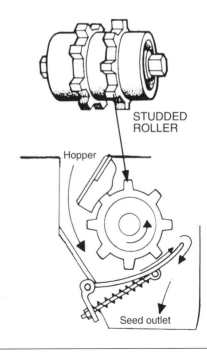

Figure 11.2   Studded roller force feed mechanism.

Coulters

Drill coulters are carried on arms which pivot individually at the front of the drill frame. Springs are used to hold the coulters at the correct depth in the soil. Spring pressure can be adjusted to suit soil conditions, sowing depth and different types of soil, with hard land requiring the most pressure. Some minimal tillage and grain-only drills have hydraulic rams to increase or decrease the working depth of the coulters. A regular sowing depth is important for the even emergence of the seedlings.

Suffolk coulters have an iron or steel shoe which cuts a narrow furrow for the seed. Its shape helps to keep a straight and even-depth furrow. Suffolk coulters do not block easily and are not affected by stony land. There are no moving parts to lubricate, but the bottom of the coulter will wear away quite quickly in some abrasive soils. Individual spring pressure on the coulter units is used to hold the coulters at the required depth, but penetration can be difficult in hard ground. Some Suffolk coulters have a hinged flap (see figure 11.5) to prevent soil blocking the outlet if the drill is reversed while they are in the ground. Suffolk coulters are suitable for sowing both cereal and root crops.

Plate 11.10   Individual depth wheels are used to control the working depth of these double-disc coulters.

Disc coulters are saucer-shaped and are suitable for drilling most of the usual arable crops, but they cut a rather uneven-depth furrow and are not suitable for root crops. Their cutting action aids penetration and they work well in most soils. Seed is dropped into the shadow of a single-angled disc, or between the pair of discs on a double-disc.

Tine or hoe coulters' main use is on cultivator/drill combinations for minimal seedbed work. Hoe coulters, attached to spring-loaded or spring tine coulter arms, can have various width shares, usually with replaceable points. The coulters part the soil and make shallow furrows for the seed.

Row Width

Row widths of 180 mm (7 in.) were used for many years when drilling cereal crops, but drills with coulters spaced between 90 and 150 mm (3½–6 in.) are in current use. Some single-pass cultivator/drills with pneumatic seeding have wider row widths. One example has double-disc coulters 240 mm (9½ in.) apart.

Pneumatic Seed Drills

Pneumatic seed drills are suitable for sowing a wide range of seeds at relatively high forward speeds. Both mounted and trailed pneumatic feed drills with working widths of 2.5 to 12 m are in common use.

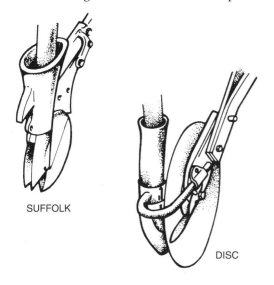

SUFFOLK

DISC

Figure 11.3   Types of drill coulter.

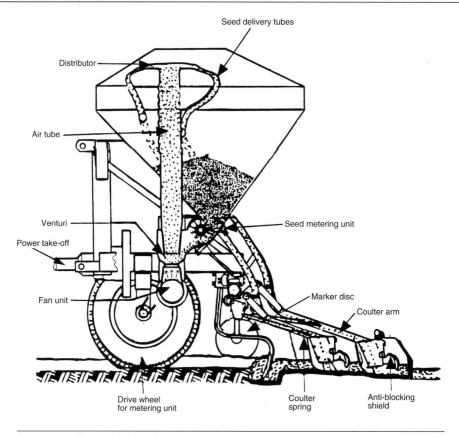

Figure 11.4   The path of seed in a pneumatic-feed grain drill.

Although the seed metering mechanism, usually of the fluted roller design, is land wheel-driven on many drills, on some it is driven by a hydraulic or an electric motor. Seed is supplied from the hopper at a controlled rate to a venturi, where it enters an air stream from a fan driven either by the power take-off or a hydraulic motor and is then blown to the distributor head. This distributes an equal amount of seed to each of the seed tubes, which deliver it to the coulters. Hydraulic rams are used to control sowing depth. Older, wide pneumatic feed grain drills sometimes have two hoppers with separate metering units that supply seed to the air distribution system, but longer hoppers are now more common on high output models.

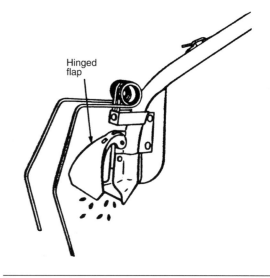

Figure 11.5   A Suffolk coulter with a hinged flap which prevents the soil blocking the coulter if the tractor is reversed when it is still in the ground.

Seed rate is altered by adjusting the metering unit with a hand lever, and the air blast fan must run at the recommended speed for efficient operation. Some drills have an electronic control box which can be used to vary seed rate while on the move. Yield maps, produced at harvest time by a combine linked to GPS satellites (see page 213), can be uploaded to the control terminal on drills with the necessary technology to vary the seed rate in different parts of the field to match its potential crop yield. The control system may also monitor seed level in the hopper, airflow from the fan and the drive to the seed metering mechanism.

Markers

Most drills have markers which can be set to make a guide mark to enable the tractor driver to make an accurate join when drilling the adjacent bout. Tractors equipped with a GPS satellite guidance system can be programmed to accurately join each bout as the tractor works its way across the field.

Drill wheelings are sometimes used when using a narrow-width drill to help the driver achieve accurate joins between bouts, but markers provide a more accurate guide. Drill markers have a disc or tine carried on an arm hinged to each side of the drill. With a few older

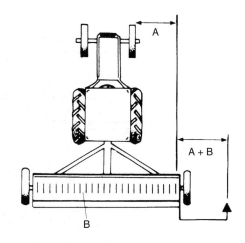

Figure 11.6   Setting drill markers.

exceptions, where a rope or lever is used, the marker arms are raised and lowered hydraulically. The marker arm is telescopic, so that the distance between the marker disc or tine and the coulter at the outer end of the marker arm can be altered to suit the tractor wheel width.

To set drill markers, refer to Figure 11.6 and follow this method:

1. Measure the distance from the centre of the tractor front wheel to the outside coulter on the drill (distance A).
2. Measure the row spacing (distance B).
3. Add A to B and position the marker disc or tine at this distance from the outer drill coulter on both marker arms.

Tramlines

These are unsown rows which provide wheelings for spraying and spreading fertiliser after crop emergence. This technique, used on most farms, avoids the tractor wheels damaging the growing crop and ensures accurate joining of bouts. The most common tramline widths are at intervals of between 12 to 36 m and are chosen to match the working width of the farm sprayer and fertiliser spreader. The tramline width used will be a multiple of 3, 4 or 4.8 m, which are the more common grain drill sowing widths.

Plate 11.11   The markers on this pneumatic feed grain drill have been set at the correct distance from the last row drilled to ensure an accurate join between bouts. (Claydon)

There are two basic types of tramlining equipment. Earlier systems with manual controls relied on the tractor driver to remember to switch off the feed mechanisms to the two rows forming each tramline on the correct bout. Modern automatic tramlining systems are operated electronically and automatically with the drill working width and the required tramline recorded in a control box in the tractor cab. The control unit, connected to the automatic marker changeover mechanism, receives a signal when the markers are raised at the end of each bout. On completion of the required number of bouts, a solenoid shuts off the feed mechanisms to the rows which form the tramlines. With pneumatic grain drills, the solenoid shuts off the relevant seed tubes to the rows used to form the tramlines. An over-ride switch is provided for use when the automatic markers are raised to clear a blocked coulter during a bout.

Modern GPS systems mean that instead of using tramlines, some farmers use the precise bout-matching ability provided by GPS which indicates exactly where to drive the tractor when spraying or spreading fertiliser. The accuracy of the GPS system ensures the tractor runs in the same path for spreading and spraying operations throughout the season.

USING DRILLS

Most mounted and trailed drills and cultivator/drill combinations with working widths of about 3 m are narrow enough to be towed or carried on the tractor hydraulic linkage or from field to field. Wider drills are either moved along the road with an end-towing kit or, now more often, the wing sections are folded with hydraulic rams. An end-towing kit has a pair of hydraulically lowered wheels at one side of the drill and provision to attach a drawbar at the opposite side.

Most seed grain is delivered to the farm in big bags weighing 500 or 600 kg. A tractor front-end loader or a telescopic forklift is used

Plate 11.12   Hydraulic rams fold the side wings on this 6 m wide cultivator drill for transport. (Kuhn)

to carry the bags from a trailer and lift them over the drill to fill the hopper.

Some farmers drill three or four bouts around the headland before drilling the main part of the field. This ensures the coulters penetrate to the correct depth before the headlands are compacted when turning at the end of each bout. This system also provides a guide mark for lifting the drill into and out of work. The headlands may be drilled last when using a combination cultivator drill or a direct drill.

CALIBRATION

The seed rate or amount of seed sown per hectare can be calibrated, or checked, in the farmyard or in the field. Some drills are linked to a computer monitoring system in the tractor cab, which can be used to monitor the seed rate while drilling is in progress and, if required, vary the seed rate to suit the soil condition or potential yield in various parts of the field.

Settings for a wide range of crops and seed rates are given in the operator's manual. When the drill is set to the required seeding rate, a static calibration test can be made to check that the required quantity of seed will be sown at this setting.

Calibration of fluted roller and studded roller force feed drills with small diameter

Plate 11.13 The dark-coloured calibration trays at the back of the grain hopper are folded down to collect the grain when calibrating the drill by checking the application rate by hand. (Kuhn)

land-contact driving wheels, as well as pneumatic grain drills with a mechanical metering system, can be done in the farmyard. A crank handle is used to turn the feed mechanism or metering unit for a calculated number of times, usually equivalent to 1/25 hectare or 1/10 acre. The number of turns varies with the model of drill and its sowing width, and this information is given in the operator's manual.

After setting the drill as advised in the operator's manual to achieve the required seed rate, the crank handle is turned the stated number of times with the drill in gear. The grain collected in the calibration trays is weighed, and multiplying the kilogram weight by 25 gives the seed rate in kg per hectare (or multiply the weight in pounds by 10 to find the seed rate in lbs per acre). The feed mechanism on very wide drills is often in two sections with a separate drive to each half. Both must be calibrated and the contents weighed to complete the process. The same procedure is used for both mechanical and pneumatic feed drills.

Drills with the feed mechanism driven by a hydraulic or electric motor and linked to an in-cab computer can be calibrated from the driving seat. When linked to a speed check system, the computer will maintain the required seed

rate. If it detects a change in forward speed, the computer automatically alters the speed of the motor driving the feed mechanism to maintain the correct seed rate. Some drills have a control unit, linked to global positioning satellites, which operates a cut-off and re-engagement of the feed mechanism at the headland and automatically adjusts the seed rate according to the requirements dictated by the yield map stored on the in-cab electronic control terminal.

Maintenance of Grain Drills
Check that the coulters, feed tubes and feed mechanisms are in good condition and that the drive belts and chains are at the correct tension.

Fertiliser is very corrosive, and the fertiliser on combine drills should be cleaned out at the end of each day's work.

Thoroughly clean the drill before storage at the end of the season. Protect bright parts from rust and replace any worn or broken parts to prepare the drill for the next drilling season before storing it under cover.

PRECISION DRILL SEEDER UNITS

Mainly used for root crops and vegetables, precision seeder units have a feed mechanism

Plate 11.14 A master wheel drive gearbox transmits power to the vee-belts which drive the belt feed seeder units on this tractor-mounted toolbar.

which can be set to drill the seed at the required spacing in the row. A number of seeder units, usually between two and twelve, are attached at the required row spacing to a tractor toolbar.

A precision drill consists of a frame on two wheels with a number of mechanical- or vacuum-feed mechanism seeder units, each with a seed hopper, a coulter, a covering device and a rear press wheel. The coulters may either be single-line for sugar beet and other crops, or multi-line, which make up to four closely spaced furrows. The press wheel may have a smooth, concave or rubber-covered surface or a cage wheel.

The feed mechanism may be driven in one of the following ways:

1. Individual drive to the seed belt from either the seeder front or rear wheel.
2. Master wheel drive with the toolbar land wheels driving the seed mechanisms through a gearbox and a system of chains or vee-belts.

3. Individual electronically controlled electric motor drive to each unit. Varying the speed of the electric motors changes the seed rate and seed spacing in the row.
4. Vacuum-feed drills usually have master wheel drive for the seed wheels and a power take-off driven fan.

Precision Seeder Unit Feed Mechanisms

Belt feed A small endless rubber belt with holes punched in it carries the seed from the hopper to the outlet point. A repeller wheel, turning in the opposite direction to the belt, helps to ensure that only one seed is carried in each hole. The belt is stretched slightly at the outlet point and the seeds drop through the hole into a shallow furrow made by the coulter.

Seed rate and spacing in the row are altered by:

1. Changing the seed belts. A range of belts is available with holes punched at different intervals to change seed spacing in the row.

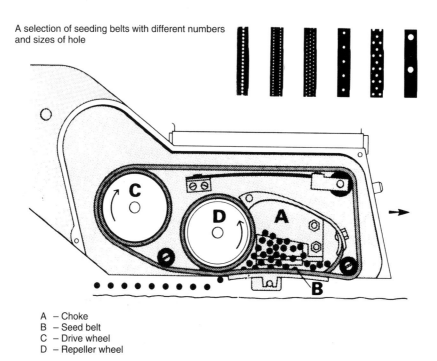

A selection of seeding belts with different numbers and sizes of hole

A – Choke
B – Seed belt
C – Drive wheel
D – Repeller wheel

Figure 11.7   Belt feed seeder mechanism. (Stanhay)

Belts with different-size holes are used to suit a variety of seeds.

2. A gearbox on precision seeders with master wheel drive provides a range of belt speeds. Only one set of belts is needed to give a wide range of seed rates and seed spacings for a particular size of seed.

Cell wheel feed Holes in the rim of the cell wheel collect single seeds from the hopper. A repeller wheel sweeps excess seeds away so that only one seed is carried in each hole. The cell wheel carries the seeds to the outlet point, where an ejector plate prises them from the holes and they fall a short distance into the soil. Seed

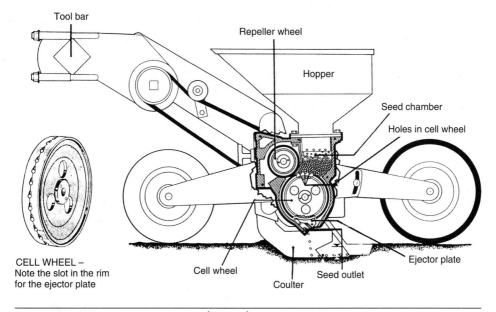

Figure 11.8   Cell wheel feed mechanism. (Stanhay)

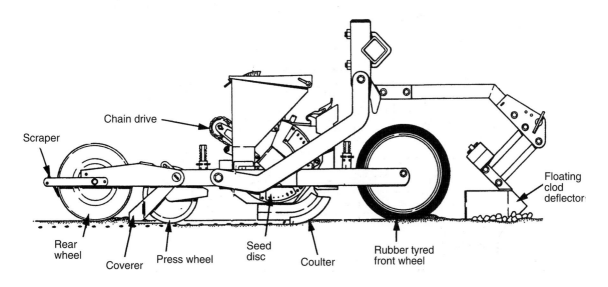

Figure 11.9   Vacuum-feed mechanism. (Stanhay)

rate and spacing are changed in the same way as belt feed. The cell wheels are changed as required for different spacings, and different-sized holes cater for various sizes of seed. A gearbox is used to vary seed spacing in the row on cell wheel seeder units with master wheel drive.

Vacuum feed This type of precision seeder has a power take-off driven fan on the toolbar which creates a vacuum in each seeder unit. The feed mechanism consists of a flexible disc, with regularly spaced holes, which turns beneath the seed hopper. The vacuum holds single seeds against the holes in the disc, which carries them to a discharge slot above the coulter. Here a blanking plate cuts off the vacuum and the single seeds fall off the disc into the soil. Airflow from the fan prevents a build-up of dust in the seed disc housing and cleans the seed-carrying holes. Changing the speed of the feed disc with the master wheel drive gearbox alters seed spacing.

Flexible seed discs with one, two, three or four circles of holes can be used to drill seed in single rows, closely spaced double rows or in triple rows for such crops as carrots. Vacuum-feed seeder units are suitable for many types of seed including celery, parsnip, parsley and cabbage.

Plate 11.16   Sugar beet, oilseed rape, maize and sunflowers can be drilled with this precision seeder at row spacings of between 450 mm and 1 m. It can also be used with a fertiliser placement attachment. (Vaderstad)

Seeder Unit Performance Monitor

Precision seeders with a performance monitor have a display screen with warning lights in the cab. It monitors the rotation of the seed belts, cell wheels or seed discs to inform the tractor driver that the units are working correctly. A separate sensor alerts the driver when the seed hoppers need re-filling. It is usual for the hopper with the sensor to have slightly less seed than the others when it is filled to ensure that none of the other hoppers are empty before a light flashes or a warning appears on the screen. Some monitoring systems also have an audible warning, which sounds when the seed hoppers are nearly empty or the seeding mechanism develops a fault.

Using Precision Seeders

Before attaching a precision drill to a tractor, the wheel track should be set so that the wheels do not run directly in front of a seeder unit and compact the soil. The wheel track is usually set at a multiple of the row width of the crop being drilled. The tractor wheels would, for example, be set at 1.5 m (60 in.) when drilling sugar beet in rows spaced at 500 mm (20 in.).

Plate 11.15   This nine-row precision seeder unit has an air-feed seed mechanism. (Kuhn)

Marker setting is calculated in the same way as for a grain drill (see Figure 11.5). However, when using a very narrow drill, if the total sowing width is less than the wheel track, then distance B is subtracted from distance A. When the tractor is equipped with GPS, the system can be used to achieve accurate joins between bouts, especially when using very wide drill toolbars. Accurate joins between bouts are important for any inter-row work and harvesting.

The headlands are normally drilled first, and it is usual to drill the sugar beet crop in multiples of the rows lifted at a time by the harvester. A toolbar with six, twelve or eighteen seeder units would be used for a crop to be lifted with a six-row harvester.

For road transport, most precision seeder toolbars more than 3 m wide are hydraulically folded, but older machines may be end-towed in a similar way to older types of wide grain drills.

CHAPTER 12

Manure and Fertiliser Distribution Machinery

FARMYARD MANURE SPREADERS

There are three types of manure spreader: rear delivery machines used mainly to spread solid manure, rotary and side spreading machines that can be used for solid and semi-liquid material, and slurry tankers which are used to spread liquid manure.

The rear delivery manure spreader is basically a trailer with a combined shredding and spreading mechanism at the rear supplied with manure by either a power-driven moving floor conveyor or a hydraulically operated pusher board. Side spreaders have a body with sloping sides and a large auger which conveys the manure to a paddle-type spreading mechanism at one side of the machine. Rotary manure

spreaders have a power take-off driven rotor shaft with a number of flails on heavy-duty chains in a cylindrical-shaped hopper. The chains throw the manure from the side of the spreader on to the ground. Flail spreaders with one or more rows of high-speed rotating flails across the back of the machine which shred and spread the manure are now less common, but are sometimes found in livestock farming areas. Many modern manure spreaders have hydraulic brakes, and the larger models have tandem sprung axles with flotation tyres.

Rear Delivery Spreaders

The heavy-duty steel floor has an endless chain and slat conveyor or pusher board which carries the manure to the rear of the spreader. Here, high-speed shredders or beaters tease out the manure and, depending on the size of the machine, spread it over a width of 6–14 m (20–45 ft). High capacity models can have spreading widths of up to 24 m. Some rear delivery spreaders have a hydraulic pusher board, operated by a multi-stage ram, which gradually pushes the manure to the spreading mechanism.

Various types of spreading mechanism are used. Some machines have a single horizontal rotor, which incorporate tines or an auger with left- and right-hand flights to shred and spread the manure. Other models have two or three horizontal shredding and spreading rotors. Some manure spreaders have two, three or four vertical spreading rotors spaced across the back of the machine or either one or two

Plate 12.1 This 10-tonne manure spreader with twin vertical rotors spreads manure at a rate of two tonnes a minute. (Teagle)

Plate 12.2   Two vertical rotors shred and spread manure carried to them by a chain and slat floor conveyor. The rear gate, which is used to restrict the flow of chicken litter and other loose material, is in the raised position. (Joskin)

horizontal spinning discs which spread the manure is a similar way to a spinning disc fertiliser broadcaster.

The moving floor conveyor and spreading mechanism may be driven by the power take-off and protected with an overload slip clutch or by a variable speed hydraulic motor supplied with oil from the tractor hydraulic system.

Rear delivery spreaders are best suited to manure with a high straw content, but some models are adapted for use with semi-liquid manure and chicken litter. A rear gate or slurry door, raised and lowered hydraulically, restricts the flow of material to the spreading rotors. A special chicken litter attachment can be used on some models of spreader with horizontal spreading rotors. It consists of a row of spinning discs, usually four in number, under a large metal cover. The rotor feeds the litter at a controlled rate to the hydraulic motor driven disc, which spreaders it over a wide area (see plate 12.4).

Plate 12.3   Rear view of a horizontal beater manure spreader with a chain and slat floor conveyor.

Capacities usually range from 5 to 40 tonnes of farmyard manure. Some spreaders have a load cell which measures the weight of the load. Application rates will vary from about 1 tonne up to 40 tonnes per hectare (20 tons/acre).

Plate 12.4 Spreading widths of 16 to 20 m can be achieved with this 12-tonne spreader, which is suitable for most types of farmyard manure including chicken litter and slurry. High-speed flails break up the material, which is spread from the rear by two spinning discs. (HiSpec)

Plate 12.5 Side spreaders can be used to distribute a wide range of manure from solid material to slurry. (HiSpec)

When loading a rear delivery manure spreader, filling should start at the front of the machine, and it will be easier to manoeuvre the tractor and loader or materials handler if the spreader is set at an angle of about 30 degrees to the heap.

Controls

Drive engagement The power take-off lever or switch on the tractor engages the drive to the floor conveyor and spreading mechanism of power take-off driven models. An auxiliary services spool valve is used for spreaders powered by a hydraulic motor.

Application rate is altered by either changing the speed of the moving floor conveyor with a lever or screw adjuster on power take-off driven machines or by varying the oil flow on spreaders with a variable speed hydraulic motor. Spreaders with a pusher board, which moves the manure to the spreading mechanism, have a variable speed hydraulic ram which varies the speed of the pusher board. Forward speed, changed with the tractor gearbox, also alters the application rate. A low forward speed and high floor conveyor speed gives a high application rate.

Rotary Spreaders

The open-sided cylinder-shaped trailer body or drum on a rotary farmyard manure spreader is also suitable for handling wet material and chicken litter. A number of high-speed flails attached to a centrally mounted rotor by heavy chains throw the manure from the side of the spreader onto the ground. The rotor shaft is driven by chain and sprocket from the power take-off. Rotor speed is varied by using different-size sprockets in the chain drive, which is protected by an overload safety clutch.

Before filling the spreader, the flails should be wrapped around the rotor. When spreading commences with a full load, the flails at each end of the rotor distribute the manure, and as the load is discharged, more of the flails are able to swing free and gradually empty the spreader from both ends to the centre. A hinged lid above the drum, operated with either a hand lever or a hydraulic ram, can be opened to facilitate filling.

Rotary manure spreaders are made in various sizes, with capacities from 3–12 tonnes or approximately 1,220–4,800 litres and a spreading width of 3–12 m depending on the size of the machine. Application rates are within a

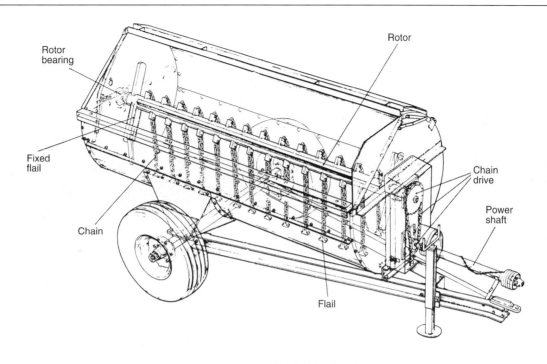

Figure 12.1   Sectional diagram of a rotary manure spreader.

Plate 12.6   Depending on the type of manure, a rotary side spreader has a spreading width of 15 m or more. (HiSpec)

range of 12–60 tonnes per hectare (5–25 tons per acre) depending on model, but this can be increased by applying a second dressing.

Controls

Drive engagement The tractor power take-off shaft or switch is used to engage the drive to the flail rotor.

Application rate is varied by changing gear to alter the forward speed of the tractor. Altering the speed of the flail rotor varies the quantity of manure spread per hectare. When the selection of different forward speeds does not give the required dressing, fitting different-sized sprockets on the chain drive to the rotor shaft will change the flail speed. Flail rotor speeds vary between 200 and 350 rpm depending on model.

Side Spreaders

An auger running in the bottom of a triangular section hopper carries manure to a paddle-bladed rotor which throws the manure sideways from the spreader. Some machines have the spreading rotor near the front; others have a centrally mounted rotor. The auger has left- and right-handed flights which carry the manure to the spreading mechanism, from where it is thrown up to 15 m or more. The auger and spreading rotor drives have shear bolts to protect them from damage if either mechanism is overloaded. The adjustable gate, which controls the flow of manure to the spreading unit, is opened and closed with a hydraulic ram.

Sidespreaders require a tractor of at least 60 kW (80 hp). They are multi-purpose machines and can be used to spread farmyard

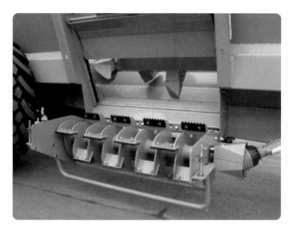

Plate 12.9 The power take-off driven spreading rotor is protected from overload by a shear bolt. The shutter above the rotor, which controls application rate, is raised and lowered by a hydraulic ram. (Shelbourne Reynolds)

Plate 12.7 This side spreader holds about 11 tonnes of farmyard manure. (Shelbourne Reynolds)

manure, slurry or chicken litter, with a capacity of up to 15 tonnes of manure or approximately 13,000 litres (2,800 gallons) of slurry.

Application rate is changed by selecting a different gear to alter the tractor forward speed, changing the speed of the main auger and by opening or closing the adjustable shutter to control the flow of material from the hopper. The shutter setting will also depend on the type of material being spread.

Using Manure Spreaders

The best results with a rear delivery spreader are obtained by working in strips or lands across the field. This will avoid undue running on the headland and making tight turns, which disturb the field surface and can result in damage to the power take-off shaft. With side-delivery spreaders, it is usual to start at one side of the field and work towards the opposite hedge. The spreader can either be filled with a materials handler, a second tractor and loader or, if the towing tractor is equipped with a front-end loader, it can, after uncoupling the spreader, be used to refill the machine.

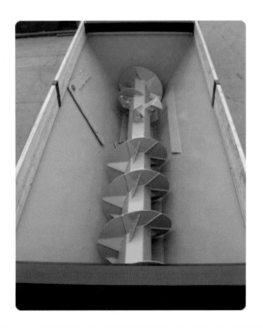

Plate 12.8 Left- and right-handed augers, turning at 750 rpm, carry the manure to the spreading rotor on this side spreader. (Shelbourne Reynolds)

Spreader Maintenance

Except for sealed bearings, any other lubrication points should be greased daily when the machine is in use. Gear box oil levels should be checked at frequent intervals during the spreading season and chain drives kept correctly tensioned. Check the condition of hoses to hydraulic rams and motors. At the end of the season the spreader should be cleaned down and damaged parts replaced so that it will be in full working order when it is next used.

SLURRY SPREADERS

Slurry can be spread with specialist irrigation equipment using a pump, pipeline and slurry gun which sprays the liquid on to the land. However, most livestock farms have a trailed slurry spreader with a cylindrical tank with capacities in the 800–4,000 litre bracket. Larger models with tandem axles and tank capacities of up to 25,000 litres are mainly used by agricultural contractors and on some large livestock farms. Environmental considerations have resulted in the increasing use of injection equipment attached to the back of the tanker, which places the slurry in slots cut into the soil.

There are several types of slurry tanker, which, depending on size, may have one, two or three axles. Many are filled with a vacuum pump on the machine, some have a mechanical filling system, while others are top-filled with a slurry pump, which is usually part of a handling and mixing system at the slurry store.

Vacuum tankers have a high capacity power take-off driven vacuum pump, which draws the liquid through a pipe placed in the slurry storage tank, pit or lagoon. When the tank is full, the pump is used to pressurise the contents to about 3 bar (45 psi). This pressure is used to expel slurry from the tank and apply it from a rear discharge spout or slurry gun, dribble bar, trailing shoes or an injector system.

Tankers with a rear discharge spout (which due to the odour they disperse are no longer environmentally fashionable) are still used on some farms. The slurry is pumped against a deflector plate which spreads it in a band up to 20 m wide. Slurry spreaders with rear-mounted dribble bars, with typical working widths from 7 to 24 m, are folded vertically to give an overall transport width of about 3 m. A typical dribble bar slurry spreader has 40 mm diameter flexible tubes at 350 mm spacing that place the slurry in bands on the surface of the

Plate 12.10   Spreading widths from 11 to 15 m are possible with this vacuum slurry tanker. (HiSpec)

Plate 12.11   The discharge spout on this slurry spreader has a low profile spread pattern. (HiSpec)

Plate 12.12   A dribble bar attachment with spreading widths of 7.5 to 24 m can be used with this slurry tanker. (HiSpec)

Plate 12.13   The thirty-four outlets on this trailing shoe boom have a 7.5 m spreading width. (HiSpec)

Plate 12.15   The individually spring-loaded tines on this rear-mounted toolbar cut slots in the ground for the injector nozzles to do their work. (Joskin)

Plate 12.14   Slurry is injected into the slots cut by a double row of tines on the toolbar attached to this slurry tanker. (Joskin)

Plate 12.16   Discs are used to cut slots in the ground for injecting the slurry with this tanker. (Joskin)

field. The trailing shoe system has rigid tubes with steel coulters which place them in grooves on the surface of the soil.

Slurry injection systems, which deposit the slurry in narrow slots cut in the soil, have helped to overcome some of the complaints from the public concerning air pollution when spreading slurry. A frame attached to the back of the slurry tank has a number of heavy-duty tines or discs which cut slots in the soil; the slurry, pumped through flexible tubes attached to the tines or discs, is deposited in these slots. The injector bar, typically between 4.5 and 7.5 m wide, is folded behind the tank for transport.

Application Rate

Slurry spreaders can run over the same area several times, so output is best measured in the time taken to fill and empty the machine. A typical 4,500-litre tank can be both filled and emptied in 3–5 minutes.

Slurry Mixers

The pump on a slurry tanker is sometimes used to pump its contents back into a storage tank, pit or lagoon to break up surface crusts or mix liquids with solid matter. Mixing can also be done with a tractor-mounted mixer with a power take-off driven rotary stirrer which is lowered into an open lagoon or pit.

Plate 12.18   A slurry tanker with the tool bar folded for road transport. (Joskin)

Typical machines have a 5 or 7 m stirrer arm for mixing contents, which can be raised and lowered mechanically, or with a hydraulic ram, to stir the slurry at different levels in the pit or lagoon. Other types of mixer are used to stir the contents of slurry storage tanks.

FERTILISER BROADCASTERS

Fertiliser may be spread with a broadcaster or a full-width distributor. The twin-disc broadcaster has become the most popular machine, but single-disc broadcasters, oscillating spout broadcasters and full-width distributors are still in use.

Plate 12.17   A hydraulically operated suction pipe can be used to fill a slurry tank. (HiSpec)

Plate 12.19   The hopper on this twin-disc spinner broadcaster has a tilt mechanism for boundary spreading. Depending on model, the hopper on this spinner broadcaster has a capacity of 1,500–3,000 litres, with spreading widths of 6–36 m. (Teagle)

Spinning Disc Broadcasters

This type of mounted or trailed broadcaster has a hopper made with stainless steel, coated steel or tough plastic mounted above the spreading mechanism, which is driven either by the power take-off or a hydraulic motor. Most broadcasters are tractor mounted, and depending on model, have a hopper capacity of between 700 and 3,000 litres or more. The hopper capacity of many mounted broadcasters can be increased with the addition of one or more extension kits. One example increases the hopper capacity from 1,000 to 1,500 litres. Typical hopper capacities for trailed broadcasters range between 6,000 and 18,000 litres. Most manufacturers quote hopper capacity in litres. The litre weight of different types of fertiliser varies widely, but 1,000 litres of a typical granular type weighs about one tonne.

Most broadcasters have a screen in the hopper to prevent lumps of fertiliser or stray material clogging the outlet point above the disc(s). Some machines have an agitator in the bottom of the hopper to prevent the fertiliser bridging, especially when working in a damp atmosphere. Spinning disc broadcasters have either one or two horizontal, stainless steel discs with stainless steel or plastic vanes which spread the fertiliser evenly over the ground.

Plate 12.20   A mounted single-spinning disc broadcaster with a spreading width of up to 12 m. (Teagle)

On most spinning disc broadcasters, the power take-off should be run at 540 rpm, but always consult the operator's manual. As smaller broadcasters have a low power requirement, fuel economy can be achieved when using a tractor with a two-speed power take-off in the 1,000 rpm setting so that the power shaft runs at 540 rpm at a lower engine speed. The drive, which is protected by an overload clutch, is arranged on twin-disc machines so that the discs contra-rotate, with both either rotating towards or away from the centre of the machine. The discs are timed with each other to give the best possible spread pattern and to prevent fertiliser colliding as it is thrown off the discs.

Application rate is changed by regulating the flow of fertiliser from the hopper to the disc(s) and by changing the forward speed of the tractor. An adjustable shutter adjusted by a hand lever or a small hydraulic ram operated remotely from the cab controls the flow of fertiliser from the hopper. The range of application rates varies with different models of broadcaster. A typical machine will spread between 3 and 1,000 kg/ha. Variable application rates, determined by the data taken from the field yield map, can be achieved when using a tractor linked to global positioning satellites.

Spreading width can vary from 6 to 50 m, depending on the type of fertiliser being spread, the size of the machine and the way it has been set. The variable factors that affect the spreading width of spinner broadcasters include the type and position of the vanes on the spinning disc(s) and, on some broadcasters, discs with a different spread pattern can be used. Changing the position of where the fertiliser falls onto the disc(s) and the height of the disc(s) above the ground also affects the spreading width. When applying fertiliser to crops with tramlines, the spreading width must match the spacing of the tramlines or row widths.

The spinning disc(s) on some broadcasters are driven by a hydraulic motor that provides an infinitely variable disc speed, allowing the broadcaster to operate at different spreading widths.

The type of fertiliser also affects the spreading width. Granules, prills and powdered fertilisers vary in size, weight, density and hardness, and each of these factors determines how far the same machine will spread a particular fertiliser. Fertilisers with a high weight per litre will be thrown further than one with a low litre weight. A typical twin-disc machine has spreading widths ranging from 16 m with light fertilisers up to 24 m with heavier and denser materials.

Checking Application Rate

The high cost of fertiliser calls for the accurate distribution of it to achieve high yields and prevent striping in the crop caused by misses or overlaps when spreading. Instruction books give the settings for a wide range of fertilisers and application rates. Spinning disc broadcasters can be checked, or calibrated, in the farmyard. Some machines are supplied with a calibration kit which is used for a period of time specified in the instruction book. After collecting and weighing the fertiliser, a simple calculation is used to convert this weight into kg per hectare.

Some broadcasters have an electronic calibration system that can be used in the field without the need to catch and weigh the fertiliser. One design has weigh cells linked to a computer that monitors the flow of fertiliser from the hopper. After entering the required application rate and initial shutter setting, the computer constantly checks and maintains the correct flow of fertiliser to achieve the pre-set application rate. To compensate for slight changes in tractor forward speed, the computer, linked to GPS satellites, varies the spreading rate to match the requirements of the yield map made at harvest time.

Border Spreading

Environmental regulations include certain restrictions concerning the application of fertiliser on field boundaries, especially near watercourses, footpaths, wildlife margins and houses. Controlling the pattern of border spreading will also maximise crop yield on headlands and reduce waste of fertiliser on land outside cultivated areas.

Various border spreading control systems are used on spinning disc broadcasters. Some have a set of adjustable metal deflector fins attached to the back of the hopper that can be raised and lowered from the cab with a small hydraulic ram. When lowered, the deflector fins change the direction of the fertiliser as it leaves the spinning disc to reduce the spreading width on the field boundary side of the machine. Another design has a small hydraulic ram which tilts the hopper sideways and resets the angle of the spinning disc to reduce the width of spread on one side of the machine. The spinning discs on some broadcasters have adjustable vanes which can be altered to reduce the width of spread on one side of the machine.

Plate 12.21   A hydraulic ram on this broadcaster lowers the adjustable deflector unit into line with the spinning disc for border spreading.

Trailed Spinning Disc Broadcasters

Trailed fertiliser and lime spreaders, popular with farm contractors, have capacities of 5 to 15 tonnes or more. A chain and slat or conveyor belt on the hopper floor carries the fertiliser to one or two spinning discs at the rear. Either the tractor power take-off or a hydraulic motor is used to drive the floor conveyor and the spinning disc or discs are driven by a hydraulic motor.

Application rate is varied by the speed of the floor conveyor and by an adjustable shutter which releases fertiliser on to the spinning discs. The adjustable shutter also cuts off the flow of fertiliser when turning at the headland. Spreading widths, depending on the size of the machine, range from 15 to 30 m or more.

Oscillating Spout or Pendulum Broadcasters

Mounted broadcasters with a swinging or oscillating spout spreading mechanism are now less common on arable units, but can still be found on mixed and small arable farms. The spout, which swings through an arc, is power take-off driven through an eccentric unit.

Spreading width varies between 4 and 14 m depending on spout length, the height of the spout above the target and the litre weight of the fertiliser. There are various types of spout,

Plate 12.22   Oscillating spout fertiliser broadcasters have a spreading width of up to 15 m. (Vicon)

used for wide or narrow spreading widths, boundary spreading to prevent fertiliser falling near ditches, and band spreading. As the fertiliser is thrown with less force than from a spinning disc, the litre weight of the fertiliser has a significant effect on spread pattern. Heavy fertiliser granules will be thrown further than powders. Some swinging spout broadcasters distribute the fertiliser over a narrower width than spinning disc machines, and the reduced width of spread makes them unsuitable for top-dressing cereal crops with widely spaced tramlines.

The hopper capacity of mounted swinging spout broadcasters ranges from 250 to 1,500 kg. Trailed models have a capacity of 2 to 3 tonnes. Application rate is varied with an adjustable shutter in the hopper and by selecting a different tractor forward gear.

Using Fertiliser Broadcasters

Mounted broadcasters must be held rigid on the three-point linkage with external check chains or stabiliser bars. It is important to set the machine so that the hopper is level and the spinning disc or swinging spout is the correct height above the ground or crop. These settings should be checked in the instruction book. Failure to follow the maker's recommendations will affect the width of spread, which in turn may cause inaccurate matching between bouts.

Less fertiliser is applied at the outer sides of the spread pattern, and to achieve an even application across the field, it is usual for the distance between adjacent passes to be less than the full spreading width to give some overlap between bouts. When working in rowcrops or tramlines in cereal crops with a broadcaster it is easy to achieve the correct overlap between bouts. However, in the absence of tramlines or when working on bare land, the use of a satellite navigation system on the tractor (see page 213) gives very accurate joins between bouts. Alternatively, marker stakes can be placed in the ground at the required intervals across the field to achieve accurate matching between bouts.

A high lift attachment for the tractor three-point linkage, available for some broadcasters, increases the discharge height of the spinning disc(s) or swinging spout and is sometimes used for top-dressing cereals and other tall crops. Some fertiliser broadcasters have a hydraulically operated bag lifter, attached to the machine, but it is more usual to use a telescopic forklift to fill the hopper with big bags of fertiliser.

FULL-WIDTH DISTRIBUTORS

Unlike broadcasters, these distributors have an overall width very similar to their spreading width. Although most farmers spread their fertiliser with a broadcaster, some use a full-width machine with either a mechanical or pneumatic distribution mechanism.

Pneumatic Distributors

Pneumatic distributors meter the fertiliser with an adjustable roller feed mechanism into a high-speed air stream from a power take-off driven fan. The roller feed mechanism is similar to that used on a pneumatic feed grain drill. Once in the air stream, the fertiliser is

Plate 12.23 Feed rollers, driven by hydraulic motors, meter fertiliser from the hopper into an air stream, created by two fans; this carries the fertiliser to a series of outlets across the full width of the 36 m wide boom. (Kuhn Rauch)

carried along to outlet pipes or tubes with spreader plates spaced at intervals along folding booms.

A typical pneumatic spreader can have folding booms up to 36 m wide to match the more popular tramline spacings used in combinable crops. One design has a number of equally spaced spreader plates across the boom. Another has widely spaced distributor heads with each one spreading fertiliser over a wider area.

Pneumatic fertiliser spreaders, like crop sprayers, have booms which are folded either with hydraulic rams or manually for transport. Break-back mechanisms in the mounting brackets allow the booms to swing backwards if they hit an obstruction. One half of the boom can be shut off when working near or on headlands or when only a narrow strip is left to complete the field.

Hopper capacity ranges from 1,000 to 6,000 kg or more depending on model, and the hopper is large enough to be filled from big bags suspended over it by a tractor loader or telescopic handler. A mesh screen in the hopper prevents lumpy fertiliser entering the metering mechanism.

Application rate is varied by changing the speed of the feed roller through a gearbox or shutting off part of its length. One model of pneumatic spreader has a gearbox which provides 80 different feed roller speeds. Forward speed, altered with the tractor gearbox, will also affect application rate, which on a typical machine ranges from 40 to 1,600 kg per hectare for fertiliser, down to as little as 5 kg per hectare when broadcasting seeds.

Using Pneumatic Distributors

Pneumatic fertiliser distributors are calibrated in a similar way to broadcasters, and most models are supplied with a calibration bag for collecting fertiliser from the metering unit. Tramlines simplify the operation of full-width spreaders. Matching bouts is equally easy when using a tractor with GPS, but in other situations marker stakes are required to ensure accurate joins between bouts.

Electronic Monitoring Systems

The high price of fertiliser has resulted in the introduction of remote controls and electronic monitoring systems in the tractor cab to check and maintain accurate spreader performance. An uneven application rate is caused due to the spread pattern being either too wide or too narrow, or bouts being matched incorrectly, resulting in striping in the crop and reduced yields.

The performance of a fertiliser broadcaster can be checked with an electronic monitoring and control system linked to GPS with a digital display screen in the tractor cab. Information provided on the screen includes the level of fertiliser in the hopper, tractor forward speed, area covered and the speed of the spinning disc(s). Some systems monitor the wind speed and direction in the area surrounding the spreading discs. The unit may also have a remote control system for shutting off sections of the boom on pneumatic spreaders. On some models, the driver can increase or decrease the application rate while on the move with an electronically controlled actuator ram, which varies the flow of fertiliser to the spinning disc(s). More complex control systems automatically increase or decrease application rate while spreading fertiliser to match variations in forward speed and thus maintain the pre-set application rate at all times. Variations in tractor speed may be due to driving up, down or across an incline, wheelslip or changes in field conditions.

By entering details of forward speed, spinning disc speed and the required application rate in the spreader's electronic control terminal, the settings for various application rates can be calibrated and stored. The system then automatically controls the flow of material from the hopper to maintain a constant application rate irrespective of tractor forward speed. A further refinement includes a load cell on the hopper frame that measures the weight of fertiliser in the hopper and converts this information into the distance the tractor can travel before the hopper is empty.

These advanced monitoring systems can be used in conjunction with a yield map produced by a combine harvester to vary the amount of fertiliser applied to match the potential yield of any part of the mapped field. When the spreader control terminal is linked to global positioning satellites, previously also used to make a yield map at harvest time, the application rate can be automatically increased or decreased according to a pre-determined application rate stored in the computer memory.

Maintenance

Daily lubrication is important; all drives, especially vee-belts, should be checked and correctly tensioned. Careful cleaning after use is essential to ensure a long working life, and all traces of fertiliser should be removed to prevent corrosion. When a hose or pressure washer is used to clean the machine, make sure no water is left in the spreader before it is stored for any length of time. The use of plastic materials in the manufacture of fertiliser spreaders has considerably reduced the corrosion problem.

CHAPTER 13

Ground Crop Sprayers

Crop sprayers are used to apply various crop protection products, including insecticides, herbicides and fungicides, diluted with water, to the soil or growing crops for the control of insects, weeds and fungal diseases. They are also used to apply liquid fertilisers and plant growth regulators.

Crop sprayers, which may be mounted, trailed or self-propelled, consist basically of a tank, a pump and a set of nozzles to apply the spray chemical. The nozzles are attached at equal spacings along a boom or spraybar usually supported by a lattice framework. They vary in width from 12–48 m and have hydraulic, air-assisted or atomiser type nozzles. Most arable farmers use a sprayer with a 24 or 36 m boom. However, boom widths will vary depending on the size of the farm, and most importantly, the width of the drill used to sow the crop and the distance between the tramlines for use later with a sprayer or fertiliser spreader. Spraying and tramline widths must match and are usually a multiple of three times the drilling width.

SELF–PROPELLED SPRAYERS

With tank capacities ranging from 2,000 to 7,000 litres or more, self-propelled crop sprayers are used on many arable farms and by farming contractors. Many self-propelled sprayers have an air or rubber sprung suspension system and four-wheel steering to minimise crop damage when turning; this is switched to two-wheel steering when driving on the road. The transmission may either be hydrostatic or mechanical, and engines range from 90–185 kW (120–250 hp) depending on the size of the machine. Self-propelled models with wide tyres have the advantage of low ground pressure, which minimises soil compaction. Some sprayers have automatic track width adjustment to straddle rows of different crop row widths, and a reversing camera with a screen in the cab.

A typical self-propelled sprayer has a tank capacity of 1,500–5,000 litres and a boom width of 24 m, but some machines have a boom of up to 40 m wide. Being self-propelled means that the sprayer does not tie up a tractor for the season and the sprayer is ready for use at any time. The driver, seated in a front-mounted

Plate 13.1 This self-propelled crop sprayer with a 4,000 litre tank and a 24 m boom has an electronic control system with a touch screen in the cab. (John Deere)

147

cab, has excellent visibility of the crop. Some models have high or adjustable ground clearance axles, making them ideal for spraying tall crops.

Another design of self-propelled sprayer is based on a high-speed four-wheel drive tractor with a rear load platform which is used to carry a de-mountable tank, usually holding between 3,000 to 5,000 litres of diluted chemical and a 24 or 36 m boom. The sprayer can be de-mounted and replaced with a fertiliser or lime spreader. Early in the season, when crops are short and the land is soft and wet, some farmers use a sprayer unit mounted on an all-terrain vehicle to treat crops without causing excessive soil compaction.

Plate 13.2   A de-mountable sprayer unit with a 1,000 litre front tank, a 2,000 litre rear tank and a 24 m boom. (Landquip)

High-speed self-propelled sprayers with low ground pressure flotation tyres can be used to apply as little as 75 litres/ha of diluted chemical at 24 kph (15 mph) or more with an output of up to 200 ha in a day. Lower application rates mean more acres can be sprayed in a day, but may result in inefficient coverage of the chemical.

Sprayer attachments with a battery powered electric pump, plastic tank and spray booms are also made for ATVs. They can be used for all-over spraying with standard nozzles or with a hand lance.

TRAILED SPRAYERS

Trailed crop sprayers with a tank capacity of 1,500–6,000 litres are used on many large arable farms, but the weight of larger machines can cause soil compaction. Spray boom widths generally vary between 12 and 40 m, with the chosen spraying width depending on the spacing of tramlines or the width of crop beds. The main advantages of trailed sprayers include larger tank capacity than would be possible with a mounted sprayer and simple attachment to the tractor.

Plate 13.3   This trailed sprayer with a 7,000 litre tank has an air suspension system and steering axle. (Landquip)

Plate 13.4   A steering axle is a feature of this trailed sprayer with the gull wing boom folded for transport. (Landquip)

Some trailed sprayers have a tracking system that keeps the sprayer wheels on the same track as the tractor. Electronic sensors linked to the tractor detect movement between it and the sprayer. The control unit on some models actuates the sprayer axle, while others have a steering ram on the drawbar. Both systems, which are connected to the tractor hydraulic system, keep the sprayer wheels on the same track as the tractor wheels when working in tramlines or across sloping land. The system can be over-ridden if necessary and it is locked out when the sprayer is taken on the road. An axle suspension system is another feature found on high capacity trailed sprayers.

A more advanced tracking system has hydraulic rams which automatically help the driver to steer the sprayer when making tight headland turns, and crab steering can be used to keep the sprayer wheels on the tramlines when working across slopes. If the tractor or spraying vehicle has a GPS receiver, it will record areas already sprayed and switch off the nozzles at the headland or in areas of short work such as triangular field corners. Spraying with a tractor linked to GPS satellites ensures very accurate joins between each bout.

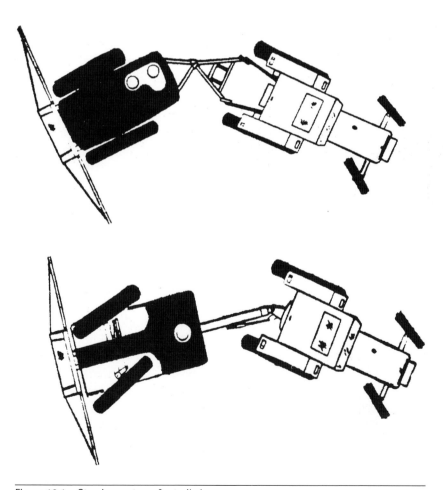

Figure 13.1   Steering systems for trailed sprayers.

MOUNTED SPRAYERS

These are made in many sizes, ranging from small basic machines with tank capacities of about 600 litres to more sophisticated sprayers with tank capacities of up to 3,000 litres. The larger models require a tractor of 150 kW (200 hp) or more, mainly due to the weight of the sprayer. Small sprayers for horticultural use with a 200 litre tank and a 6 m boom are made for tractors in the 12–15 kW (16–20 hp) range. Mounted sprayers, with boom widths of between 12 and 28 m, have an adjustable boom height which enables the driver to set it low for early season work and higher for tall crops.

Application Rates

Traditionally, sprayers have been grouped according to the amount of liquid they apply per hectare or acre. Modern chemicals are generally applied at quite low rates, and most sprayers can be used to apply the full range of farm chemicals. Low volume application is up to 250 litres per hectare (20 gallons per acre), medium volume from 250–500 litres per hectare and high volume up to about 1,200 litres per hectare (100 gallons per acre).

Very low application rates, as little as 25 litres per hectare (2 gallons per acre) have come into use in recent years, but 100 litres per hectare

is the lowest rate commonly used. These low application rates reduce the amount of water to be carried, cover more hectares per tankful and reduce the cost of chemicals. Very low application rates can be achieved with rotary atomisers instead of the normal spray nozzles. They produce uniform, medium size droplets which make very efficient use of the chemical.

There are statutory regulations concerning the application of chemicals at reduced volumes. Some chemicals can be applied at the full dose with reduced amounts of water. Others, including those classed as poisonous or toxic or where there is a risk of eye injury, must not be used at reduced volume application rates. The same regulations apply if the manufacturer states that the chemical may only be used at a reduced rate if the product dose is also reduced to match the spray volume and does not exceed the maximum concentration recommended on the chemical label.

THE PARTS OF A SPRAYER

Tank

Sprayers may have a moulded tank made of a non-corrosive plastic such as polyethylene or reinforced fibreglass; others have a stainless steel tank. Some mounted sprayers can be used in conjunction with a second front-mounted tank which holds 1,000 litres or more. The liquid in the front tank is transferred to the sprayer on the rear linkage, which reduces the time spent travelling from field to farm to re-fill. A transfer pump, usually hydraulically driven, is used to move the liquid to the main tank. It also helps to keep the contents of the front tank well mixed while spraying chemical from the rear tank. Many sprayers have a separate but integral clean water tank that holds about 10 per cent of the total volume of the main tank. After a crop has been sprayed, water in the clean water tank is pumped around the main tank and pipework for rinsing or cleaning purposes and is then sprayed out on to waste ground.

Plate 13.5   This mounted sprayer has a 1,500 litre polyethylene tank, wash tank, chemical bowl and a 12 m cross fold boom with multi-head nozzles. (Landquip)

Plate 13.6   A hydraulically driven pump transfers chemical from the front auxiliary tank to the main rear tank on this rear-mounted sprayer. (Landquip)

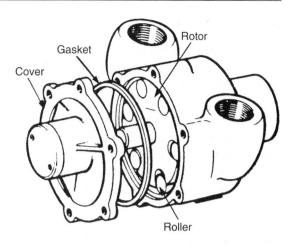

Figure 13.2   A roller vane pump.

Pump

The sprayer pump, which may be of the roller vane, diaphragm, centrifugal or piston type, is driven either from the tractor power take-off or by a hydraulic motor. Roller vane pumps are little used now because of corrosion and lubrication problems. Multi-diaphragm pumps are in common use.

The type of pump and its output will vary with the size and model of sprayer. Typical sprayer pump outputs include:

- 1,000 litre mounted sprayer with a diaphragm pump – 150 litres per minute.
- 2,000 litre trailed model with a six-cylinder diaphragm pump – 230 litres per minute.
- 2,500 litre self-propelled sprayer with a six-cylinder diaphragm pump – 300 litres per minute.
- 5,000 litre trailed model with a centrifugal pump – 700 litres per minute.

Sprayer pumps should run at the standard 540 rpm power take-off speed. Operating pressure is one of the factors determining application rate and crop penetration. Increasing the spraying pressure will reduce the size of the droplets, but small droplets will drift, even in light wind, and this must be avoided.

Roller vane pumps are relatively inexpensive to buy and maintain. With outputs of 15–20 litres/minute and operating at 2–10 bar (30–150 psi), the use of roller vane pumps is now limited to small sprayers used for weed and pest control in turf, amenity areas and some vegetable crops. A roller vane pump consists of a rotor offset to one side of its housing with a number of rubber, nylon or steel rollers (or vanes) in grooves around the outside of the rotor. The rollers are thrown outwards against the housing when the pump rotates, and liquid is drawn into the pump at the position where the rotor is furthest from the housing. The liquid is carried round between the rollers until it reaches the discharge point. Here the rotor is very close to the housing, the rollers are pushed, in turn, fully into their slots and the liquid is pumped from the outlet.

Piston pumps, with working pressures of up to 40 bar and outputs of 200 litres/minute or more, are positive displacement pumps, with the piston delivering the same volume on each stroke. The pump works in a similar way to the piston in an engine with inlet and outlet valves and piston rings to prevent loss of liquid between the piston and the cylinder wall. Using a pump with more than one cylinder increases the output. An accumulator or air bottle may be included in the sprayer circuit to reduce the pulsing effect of the piston and even out the delivery from the pump.

Piston diaphragm pumps which operate at pressures of 20 bar or more and have an output of up to 300 litres/minute are used on many current models of sprayer. These pumps may have a single diaphragm or as many as six units working in-line on high output models. The pump, shown in Figure 13.3, has a diaphragm connected to a piston which moves up and down in a cylinder. On the downward stroke, the diaphragm suction closes the outlet valve and opens the inlet valve, and liquid is drawn into the pump chamber. On the upward stroke, the diaphragm closes the inlet valve and opens the outlet valve, forcing liquid out of the chamber into the sprayer circuit.

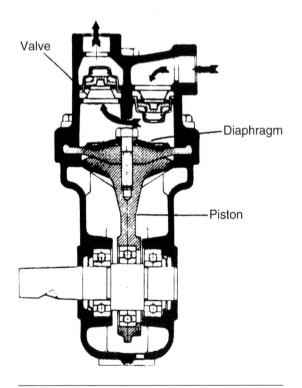

Figure 13.3   A piston diaphragm pump. (Hardi)

As with piston pumps, there can be a problem with pulsing with one- or two-cylinder piston diaphragm pumps, but this can be overcome by including an air bottle in the system. However, pulsing is not a problem with high output multiple piston diaphragm pumps, which are simple to maintain and will not be damaged if they accidentally run dry.

Centrifugal pumps with a high-speed impeller are high-pressure pumps capable of outputs of 500–1,300 litres/minute, but are sensitive to any restrictions in the system. The high output of centrifugal pumps makes them ideal for the transfer of liquid from a tractor-mounted front tank to the main sprayer tank.

Relief Valve
An adjustable relief valve is built into the sprayer circuit to vary working pressure. Spraying pressure is one of the factors that determine the application rate. The relief valve also returns excess liquid not required by the nozzles back to the tank, and agitates the tank contents to keep them well mixed.

Filters
These are a vital part of the sprayer circuit. The nozzles have very small holes that will become blocked if particles of dirt are allowed to reach them. The location and number of filters varies with different sprayers.

There are usually four stages of filtration. The first is usually a filter basket in the filler or filling system of the sprayer which filters out any dirt while filling the tank. The second is in the suction line between the tank and the pump; the third is in the delivery line from the pump to the spray boom. Final filtration is usually included in each nozzle. Sprayers with a self-fill hose have a filter unit at the end of the suction pipe. All filters must be kept clean.

Spray Boom

The spray boom or bar is made of non-corrosive material, usually stainless steel or plastic supported by a light but strong metal frame often made of aluminium. The nozzles are equally spaced along the boom. Spray boom widths of 12–36 m are in common use, but some trailed and self-propelled models have spray bars up to 40 m wide. Depending on its width, the boom will have between three and eight sections, which are folded hydraulically either vertically or alongside the sprayer for transport. The spray boom on some older sprayers is folded manually. Vertical boom adjustment is provided for different crop and target heights. Some sprayers have an automatic electronic boom height control system, with sensors that keep the boom parallel with the ground when working on slopes.

On many sprayers, the driver can turn off individual sections of the boom when less than the full spraying width is required, especially when working on triangular areas of irregular-shaped fields. With the aid of a GPS system, this can also be done automatically.

The boom is mounted on a suspension system designed to keep it parallel with the ground, reduce bounce and overcome the tendency for it to swing backwards and forwards, or yaw, on rough ground. Any unwanted boom movement will alter the position of the nozzles in relation to the spray target, which in turn can result in overdosing or missed strips. A break-back device is built into the boom hinge points. This allows the bar to swing back out of the way and reduces the risk of damage if it encounters an obstruction. Some spray booms are self-levelling when working on hillsides, with sensors that measure the distance of each end of the boom from the ground and adjust it as necessary.

Plate 13.7   Four-wheel hydrostatic transmission, a maximum road speed of 26 mph and four optional boom widths from 18 to 28 m are features of this self-propelled sprayer. (Landquip)

Plate 13.8   Multi-head nozzle holders allow a sprayer operator to select a different droplet density, spray pattern and therefore application rate simply by turning all the bodies across the boom to a different, matching nozzle, although only one blue nozzle is fitted to the four-nozzle holder shown above.

Nozzles

There are various types of spray nozzle which atomise the liquid. Most nozzles are made from plastic; others have a metal body with a ceramic tip. Plastic nozzles are colour-coded according to their output and classified by an international system as very fine, fine, medium, coarse and very coarse. Insecticides, herbicides and fungicides are applied with fine, medium or coarse nozzles depending on the intended target and stage of plant growth. The most suitable nozzle for a particular product is usually stated on the container label. Spray drift is normally not a problem when using coarse nozzles. Droplets from fine nozzles are more likely to drift, but they provide a better coverage of chemical on the leaves.

Fan nozzles with a flat fan triangular spray pattern are in common use. Fan angles can range from 80 to 110 degrees.

Cone nozzles are similar, but can produce either a hollow or solid cone spray pattern. However, the atomised chemical is prone to drift. Hollow cone nozzles are used with air-assist sprayers (see page 158) which create a curtain of air to give good spray coverage and penetration in dense crops.

Air-induction or twin-fluid nozzles use low-pressure air introduced into the nozzle to assist in the atomisation of the liquid.

Air-inclusion nozzles use a venturi system to produce larger droplets with air bubbles which reduce drift but are still retained on the leaf.

Deflector nozzles, also known as flooding or anvil nozzles, operate at low pressures and produce a flat wide angle spray pattern. They are more commonly used for liquid fertilisers and low level band spraying.

Many sprayers have a multi-head nozzle holder, which can be rotated to change the size or type of nozzle used.

Check valves, controlled by the pressure in the spray line, prevent chemical drip from the nozzles when the sprayer is turned off. They also ensure full delivery from the nozzle when spraying restarts.

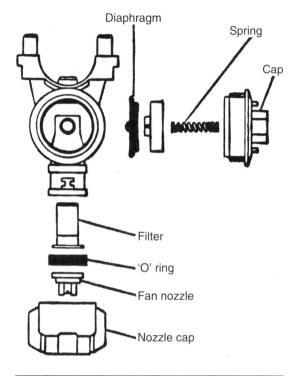

Figure 13.4　Spray nozzle with an anti-drip diaphragm valve.

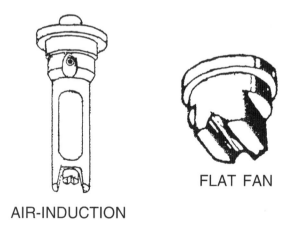

AIR-INDUCTION　　　**FLAT FAN**

Figure 13.5　Types of spray nozzle tip.

Mixing Devices

Most crop sprayers have a chemical induction bowl or a chemical induction probe

Plate 13.9 This trailed sprayer has a chemical mixing (induction) hopper incorporating a chemical can wash. (Landquip)

which is used to add the chemical to water in the sprayer tank. The chemical induction bowl, or mixing hopper, is used in conjunction with the pump and the main sprayer tank (Plate 13.9). Protective clothing must be worn while adding the correct amount of chemical to the induction hopper. After the hopper lid is closed, a valve opens and the pump sucks the contents of the induction hopper into the tank. Powders, tablets and dissolvable bags of chemical can be added after partly filling the induction tank with water, and when mixed, the contents are drawn into the tank.

A chemical induction probe is used by placing it into the chemical container, and the required quantity is drawn into a small measuring tank on the sprayer. The chemical is then transferred into the main tank and thoroughly mixed by pumping the tank contents around the system. The dilute contents of the sprayer tank are recirculated through the induction hopper to thoroughly mix the chemical as it is added, and a can wash is provided to rinse empty chemical containers.

Some sprayers have a pump-operated self-filling suction hose used to pump water from a water bowser. This overcomes the need to travel to and from the farm every time the sprayer tank is empty. A self-fill hose must not be used to fill a sprayer tank from livestock drinking troughs, ponds, streams or any water source where pollution from spray chemicals might occur.

THE SPRAYER CIRCUIT

The basic parts of a simple crop sprayer consist of a tank, pump, filters, pressure regulator and gauge, control valve, boom and nozzles. The very basic sprayer circuit shown in Figure 13.6 was used in earlier days, but modern electronically controlled sprayers are much more complicated. A typical crop sprayer has an induction hopper, clean water tank used to flush the system to remove all traces of chemical when spraying is completed, hydraulic tank agitation and controls for each section or group of nozzles along the boom.

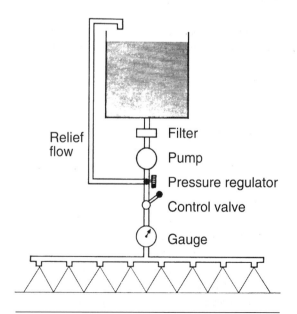

Figure 13.6   Flow diagram for a simple crop sprayer.

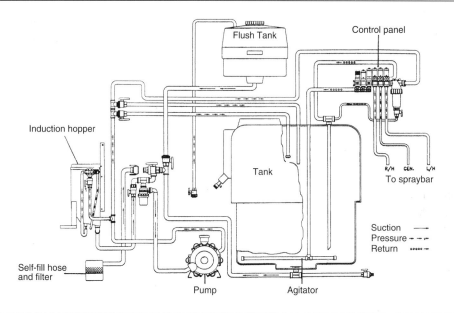

Figure 13.7   A typical flow diagram for a mounted crop sprayer with self-fill, chemical induction hopper and hydraulic agitation.

Sprayer Controls

Regulations concerning the application of crop protection products introduced in recent years have resulted in a complete change in the control and use of crop sprayers. Electric controls with computers, solenoids and electro-hydraulics have completely replaced the hand levers previously used to operate taps, valves and switches which were common on earlier crop sprayers.

The basic functions of a sprayer control unit are to switch the sprayer off and on, vary boom height, regulate spraying pressure, change the application rate and switch on or off groups of nozzles on the boom. When the sprayer is switched off, the chemical is sucked back from the nozzles to prevent it dripping from the boom, and the pump circulates the chemical around the system to keep the tank contents mixed. Each section of the boom, or group of nozzles, can be shut off when the full width of the boom is not required. Adjusting the pressure relief valve manually or with an electro-hydraulic switch regulates spray pressure.

Some sprayers have an in-cab electronic control system. A typical control unit can be used to vary spraying pressure, switch on and off the flow of liquid to the boom and control individual sections across its width. An electro-hydraulic boom tilting mechanism for keeping it level with the ground when working on slopes will be found on some sprayers. More advanced control systems adjust the height of the boom and also fold or unfold it in a set order. Other functions include locking the boom suspension system for transport and automatic tank agitation to keep the contents well mixed when not spraying.

Sprayers with an electronic control system linked to a GPS receiver provide information on a monitor screen indicating the area sprayed, the amount of chemical applied and the quantity still in the tank. When linked to GPS, the computer can carry out the headland turn procedure that includes turning off the nozzles, turning the steering system from two-wheel to four-wheel steer, matching up with the previous bout and turning the nozzles on again. On some machines, the system can also be used to shut off selected sections along the boom as triangular areas of sprayed land are passed, while on others it is possible to switch off individual nozzles.

Another electronic control system, which links multi-head nozzles to GPS (Plate 13.8), enables the operator to rotate the nozzle to choose a different nozzle type according to target height, weather conditions or the crop being treated. For example, coarser droplet nozzles can be selected when spraying a dense crop.

WORKING ADJUSTMENTS

Application Rate

Three factors affect the application rate of a crop sprayer:

1. *Tractor forward speed* A low forward speed gives a high application rate. By changing gear to double the forward speed, the application rate will be halved.
2. *Spraying pressure* Higher application rates are obtained by adjusting the relief valve to increase working pressure, or using the sprayer controller to do this automatically. High pressure reduces droplet size, and this increases the risk of spray drift with the chemical being blown away by the wind. This must be avoided, as drifting chemical can damage nearby crops and gardens. Spray drift is also a serious threat to the environment as it will harm trees, hedgerows, wild flowers, and pollute nearby water.
3. *Nozzle size* Nozzles with a large hole will give a high application rate. Smaller size nozzles are used to reduce the application rate.

Once set, the application rate will be kept constant when using a sprayer linked to an automatic electronic control system. Some machines also have a system of pressure compensation, which avoids variations in working pressure when a section of the boom is turned off.

Boom Height

The boom must be level and parallel with the ground when in work. Hydraulic rams are used to move the boom up or down on its mounting to adjust its height. This is an important adjustment as incorrect nozzle height will cause overdosing or missed strips. The boom height may be set manually on some older sprayers.

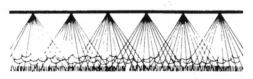

Too high

Set at the correct height, alternate nozzle spray patterns from this spray boom meet above the target so that this young cereal crop receives even coverage of chemical.

Too low

Set too low, this sprayer is leaving an untreated strip between each nozzle.

Figure 13.8   Setting spray boom height. Set at the correct height, the alternate nozzle spray patterns meet above the target so that this young cereal crop receives an even coverage of chemical. Set too low, the sprayer is leaving an untreated strip between each nozzle.

When spraying weeds that are taller than the crop, the weeds must be considered to be the target. On most sprayers, the nozzles are spaced at 500 mm along the boom. The height of the nozzles above the target is determined by the angle of the spray pattern. For example, nozzles at 500 mm spacings producing an 80 degree fan-shaped spray pattern should be about 550 mm above the target. Nozzles with a 110 degree fan pattern will need to be set lower, at about 350 to 400 mm above the target.

SLEEVE BOOM SPRAYERS

Some growers use this type of sprayer for the application of insecticides and fungicides to potatoes and other crops. Sleeve boom or downward air-assisted sprayers have a collapsible sleeve above the boom which is connected to a high capacity fan driven by a hydraulic motor. The sleeve directs a blast of air from the fan downwards and rearwards, to create a screen of air which drives chemical from the nozzles down into the crop. The air turbulence also causes increased movement and gives better penetration of the foliage being treated. Spray drift is reduced when using an air-assisted sprayer; application rates

Plate 13.10　Sleeve boom sprayers have tank capacities and spraying widths similar to conventional hydraulic nozzle machines, but use forced air to stabilise the spray droplets.

are reduced and chemical costs per hectare are much lower than with a conventional hydraulic nozzle sprayer.

TWIN-FLUID NOZZLE SPRAYERS

This type of sprayer has a special design of nozzle which is supplied with both liquid and air. The liquid is pumped in the conventional way to the nozzles, and a high volume supply of low-pressure air is provided by a compressor on the sprayer. The air and liquid are mixed within the nozzle and forced through a small hole on to a deflector plate to produce a flat fan-shaped spray pattern. Twin-fluid nozzles produce a more consistent droplet size, and there is better retention of the droplets on the leaf. Because the droplets contain air, they often collapse on impact with the foliage to create even smaller droplets and so improve coverage.

BAND SPRAYERS

Band sprayers are very similar to ordinary crop sprayers, but they apply chemical in narrow bands using a carefully calibrated and matched set of nozzles to save chemical costs where overall coverage is not required. Band sprayers are sometimes used with precision seeder units to apply pre-emergence herbicides while drilling certain vegetable crops. This allows the young plants to emerge in a weed-free band of soil to make inter-row hoeing an easier task.

Another type of band sprayer has two separate tanks, two pumps, two spray circuits and two sets of spray nozzles. One set of nozzles is used to spray a selective herbicide over the row, controlling the weeds but without harming the crop. The second set of nozzles, under dome-shaped shields, applies a non-selective herbicide between the rows to kill all weeds growing there.

USING SPRAYERS

Before the season starts, carry out the following maintenance tasks:

- Repair or replace any damaged or worn parts.
- Turn the pump by hand to check it has not seized.
- Attach the sprayer to the tractor and connect the pump to the power take-off shaft or hydraulic spool valve.
- Personal protective equipment (PPE) must be worn when carrying out any work on the sprayer which requires the operator to handle any part of the machine.
- Wash and replace the filters, fit a set of clean nozzles and then spray out a tankful of clean water, checking the nozzles for spray pattern and all parts of the machine for leaks.
- Check that all the nozzles have the same output. To do this, run the sprayer using clean water. Collect water from each nozzle with a measuring jug for thirty seconds and note the output of each nozzle. Replace any unsatisfactory nozzles. They will wear with use and, after a time, will need replacing.
- All operators must hold the relevant Pesticide Application (PA) qualifications for their machine and, under the National Sprayer Testing Scheme, crop spraying machines must have regular mechanical checks. See www.nsts.org.uk for details.

Checking Application Rate

The accurate application of spray chemicals is most important. It can be checked, or calibrated, using a static test in the farmyard or in the field. Many sprayers can be calibrated electronically using the sprayer's electronic control system. For other sprayers the following static calibration procedure can be used to check the application rate. Calibrating a crop sprayer is included in the Pesticide Application (PA) tests, (see page 161) which are required by all spraying operators.

Static calibration is carried out by running the sprayer for a calculated period of time, which is equivalent to spraying one hectare using the nozzle and spraying pressure given in the instruction book for a stated application rate. Although the calibration procedure is carried out with water, suitable protective clothing must be worn to avoid contact with any residual chemicals in the sprayer.

Certain information is required before the application rate can be checked:

1. Determine the spraying speed. If this is not known, it must be measured with the tractor in the gear that will be used for spraying and with the throttle set to give a power take-off speed of 540 rpm.
2. Mark out a distance of 100 m.
3. Drive the tractor in the selected gear over this distance and record the time taken in seconds.
4. Calculate the forward speed using this formula:

$$\text{Speed (kph)} = \frac{360}{\text{Time taken to drive 100 m}}$$

5. Measure nozzle spacing.
6. Calculate nozzle output in litres per minute by running the sprayer at the correct engine speed and setting for one minute and collect the liquid from the nozzle in a measuring jug. Carry out this test on at least four different nozzles across the width of the boom and calculate the average output from the four tests.

Using this information, sprayer output can be calculated with the following formula:

$$\text{Output in litres/hectare} = \frac{\text{Average nozzle output (litres/min)} \times 600}{\text{Speed in kph} \times \text{Nozzle spacing (m)}}$$

Example:
Spraying speed = 6 kph
Nozzle spacing = 500 mm or 0.5 m

Average nozzle output = 1.1 1itres/min

$$\text{Output l/ha} = \frac{1.1 \times 600}{6 \times 0.5} = 220 \text{ litres/ha}$$

If the output achieved in the test does not match the application rate at the settings given in the operator's manual, the test should be repeated after making a slight adjustment to spraying pressure.

Field Check

It is a simple task to check application rate against the area covered when using a sprayer with an electronic monitoring system. Using the above example, on completion of one hectare the monitor screen should record an application rate of 220 litres per hectare.

Without the advantage of an in-cab computer, field calibration is done by driving the tractor and sprayer over a calculated distance using the forward speed, nozzle size and spraying pressure to give the required application rate. Use this formula to find the distance to travel to spray 0.5 hectare:

$$\text{Distance (m)} = \frac{5{,}000}{\text{Spraying width (m)}}$$

To find the spraying width, multiply the number of nozzles by the nozzle spacing. For example, if a sprayer has 25 nozzles spaced at 0.5 m, this gives a spraying width of 25 × 0.5, which equals 12.5 m.

$$\text{Distance to travel (m) for 0.5 hectare} = \frac{5{,}000}{12.5} = 400 \text{ m}$$

With a full tank of water, drive the calculated distance, then refill the tank, measuring the amount of water needed; double this quantity to find the application rate per hectare.

SPRAYING

Set the sprayer to give the required application rate as shown in the instruction book. Fit the correct nozzles and then adjust spraying pressure to the stated level while spraying clean water. Select a forward gear to give the required forward speed with the engine revolutions at the correct speed for the power take-off drive to the sprayer pump. Read the instructions on the container of the chemical being used and wear the correct protective clothing while handling it.

Many crop sprayers, especially those with large tanks, have a chemical induction bowl (see Plate 13.9), which is a much safer method of mixing the spray chemical with the water in the tank. For sprayers without this facility, the chemical must be added with the water at the top of the tank. Partly fill the tank with water with the filter basket in position, add the chemical as instructed on the container and then fill the tank with the remaining quantity of water required to complete the mix. After the tank contents are pumped round the sprayer circuit, they will be thoroughly mixed and ready for use.

Once in the field, either spray the required number of bouts around the headland to suit the number of tramlines there or spray two full bouts around the headland before spraying the main part of the field working from one side towards the other. When tramlines are used, accurate drilling will influence accuracy of spraying. When spraying grassland or other crops without the benefit of GPS or tramlines, some farmers use marking stakes placed at the required intervals across the field as a driving guide.

SPRAYER MAINTENANCE

Most sprayers are now equipped with an integral clean water tank, usually holding 10 per cent of the total tank volume, which is used to flush out the sprayer and remove any chemical residues. When spraying is completed, water from the clean water tank can, by pressing a switch, be transferred to the main tank and circulated through the sprayer to remove all traces of chemical. The washings should be sprayed out on the same field that had been sprayed. Never dispose of the washings where they can harm humans or animals, or near a watercourse. When changing to a different chemical, also remove and clean the filters and nozzles.

At the End of the Season

Many sprayers are now used throughout the year, but if the machine is to be stored for a period, the tank and sprayer circuit should be thoroughly cleaned and flushed out with clean water. Clean and store the nozzles separately. Drain the pump to prevent frost damage.

SAFETY WHEN SPRAYING

Personal protective equipment (PPE) must be worn when cleaning and maintaining sprayers. Contaminated clothing must be stored in a PPE storage box on the machine before entering the tractor cab.

Pesticide regulations have brought about a complete change in the way that crop sprayers are operated in the field. If the driver is seated in a tractor cab with an approved air filtration system and using a sprayer with computerised electronic controls, appropriate PPE will only be required when filling, cleaning and servicing the machine. This clothing must not be worn in the cab, where it could contaminate the interior.

The current Code of Practice for the use of Plant Protection Products, and the Control of Substances Hazardous to Health (COSHH), apply to all types of farm chemicals including insecticides, herbicides, fungicides, growth regulators and soil sterilants. Hand-held sprayers, aircraft equipment as well as tractor and self-propelled sprayers are all included in these regulations.

Sprayer operators must be fully qualified to use the type of sprayer they are working with. All operators must take the necessary training and pass a Pesticide Application test PA1 to handle pesticides, together with the relevant

PA qualifications for the type of pesticide applicator they wish to use. For example, PA2 is the qualification required to apply pesticides with a boom type crop sprayer, while PA6 is applicable to hand-held and knapsack pesticide applicators.

The above information is very basic, and is only an indication of the contents of the various regulations concerning the application of pesticides. The official Code of Safe Practice is the best source of information, and you should study it to find out how the provisions affect you when handling or using crop spraying chemicals.

Some important points to remember include:

- Follow the instructions on the chemical package. Use protective clothing as directed on the label and refer to employers' risk assessment requirements.
- Check the label for any special washing requirements for the sprayer, clothing and operator that must be followed after using the product.
- Triple-rinse cans after washing out to remove any traces of undiluted chemical.
- Dispose of empty containers only through an appropriate licensed waste handling facility.
- Keep all pesticides in a locked store. Return any unused material to the store.
- Do not smoke, eat or drink while spraying and always wash before doing any of these after spraying is completed.
- Keep spare nozzles in the tractor cab. Never blow at a blocked nozzle with the mouth and don't attempt to clear a blocked nozzle with a pin either – this will damage it.
- Never transfer chemical into other containers.
- Clean all protective clothing after use.

GRANULAR APPLICATORS

Granular applicators can be used to apply crop protection chemicals in granular form to growing crops or in bands when, for example, planting potatoes and other row crops. They can also be used to apply slug pellets.

The granules are carried in a small hopper, or hoppers, and the applicator may be attached to tillage equipment, an ATV or a potato planter. Granular applicators can also be used on a combine harvester to broadcast oilseed rape seed on the harvested stubble. The flow of granules from the hopper can be adjusted to give a range of application rates.

There are two types of granular applicator: one has a single or two spinning discs driven by an electric or hydraulic motor. Depending on the size of the machine, it broadcasts granules over a width of between 12 and 36 m. The second type has a mechanical or pneumatic-fed mechanism that delivers the granules to nozzles spaced across a boom or attached to a drill and fixed to the coulters.

WEED WIPERS

Tall weeds including docks and thistles can be killed on pastures, in sugar beet and other crops with a weed wiper. This is a low cost operation compared with overall spraying. There are two types of weed wiper: one has a rope to carry the chemical and come into contact with the weeds, and the other has a wide chemical carrying roll. Both machines apply a contact herbicide to foliage tall enough to be wiped by the chemical-impregnated rope or roll.

Rope weed wipers have an endless rope wick running on pulleys attached to a front- or rear-mounted frame. Herbicide in a container is picked up by the wick and deposited on any weed it touches. Roller weed wipers work in a similar way.

Weed wipers are made in various working widths and the height of the wick or roller can be adjusted. The wider models of weed wiper can be folded for transport.

Plate 13.11 A granular applicator mounted on an ATV is used on many farms to apply slug pellets to young crops.

Plate 13.12 A roller containing herbicide coats the smooth roller on this weed wiper with chemical, which is wiped onto any weeds growing above the crop or onto tall weeds on grassland. (Carier Rollmaster)

CHAPTER 14

Haymaking Machinery

Grass is now most commonly preserved as silage for winter feed on most dairy and beef farms. It is cut and gathered while still moist and stored away from air by sealing it in plastic, either in pits or clamps or in large round or large square bales. Silage making is less weather dependent than haymaking, but the cost of production can be higher because of the use of plastic wrapping or sheets or the large concrete area required when making silage in a clamp or pit. Some of the machinery used to make both hay and silage is identical.

MOWERS

Mowers are used to cut grass and other forage crops with the mown material left in swaths for further treatment in the hay or silage making process. Many farmers speed up hay and silage making with a mower conditioner, which hastens the drying process by bruising the leaf and stems with a flail rotor or rollers so that it loses moisture more quickly.

Cutter bar mowers were used for many years, but they have fallen out of favour as they are unsuited to uneven or stony ground, and knife sharpening was an added task. Grass for making hay and grass silage crops is now usually cut with a drum or disc rotary mower, or less commonly a flail mower.

ROTARY MOWERS

Rotary mowers have a high work rate and are relatively immune to blockages, even in heavy or tangled crops. Working speeds of up to 15 kph can be achieved in good conditions.

Most rotary mowers are rear-mounted or trailed with a power take-off driven cutting mechanism. Some farmers use a front-mounted mower, leaving the rear linkage free for either a single mower, or more usually, twin rear-mounted mowers (Plate 14.2) that leave a swath on each side of that cut by the front mower.

Plate 14.1   This 3 m cut mounted rotary mower has four cutting drums fitted with knives. (Kuhn)

Plate 14.2   The front- and rear-mounted drum mowers on this tractor have a combined cutting width of 10 m. (Kuhn)

Most drum and disc rotary mowers have cutting widths from 2.8 to 3.2 m, and those with one front-mounted and two rear-mounted mowers generally have cutting widths of up 10 m. Drum mowers have two, three or four large diameter drums, each with two or three replaceable blades similar to those on a domestic rotary mower. Disc mowers have smaller diameter discs, each with two or three free swinging

Plate 14.3   Rotary conditioning tines on this four-drum rotary mower reduce the drying time for the freshly cut swath by bruising the grass. (Claas)

replaceable blades and a cutting width of about 400 mm. A disc mower has between five and nine cutting discs with a typical 3.0 m cut; a mounted disc mower will have seven discs.

DRUM MOWERS

The large diameter rotors on a drum mower are top-driven by either a system of gears or toothed belts. This positive drive system is necessary to prevent the blades on adjacent drums coming into contact with each other. The cutting rotors run at speeds in the region of 1,000 rpm.

The drive is arranged so that the drums contra-rotate on a two-drum machine to leave a single swath behind the mower. Some wider rotary mowers with four cutting drums leave two separate swaths. Trailed models often have a central drawbar which is used to offset the mower for cutting and position it behind the tractor for transport.

A typical two-drum mower with a cutting width of 1.9 m needs a tractor of about 40 kW (55 hp); a four-drum model, cutting a 3 m swath, requires at least 55 kW (75 hp). Adjustable deflectors behind the cutter bar are used to vary the width of the swath left behind the mower to aid drying and match the working width of the following machine.

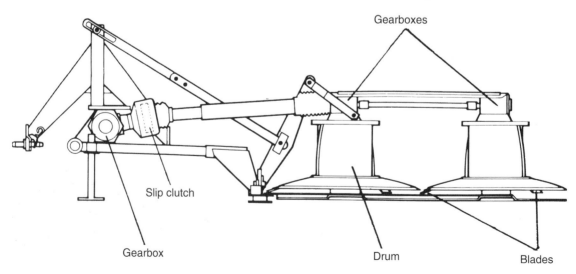

Figure 14.1   The line of drive of a typical power take-off driven drum mower.

Plate 14.4   A diagrammatic view of a cutting drum and conditioning tines on a drum mower. (Claas)

Plate 14.6   The eight discs on this rotary mower create a cutting width of 3.5 m. The cutter bar can be angled for working on sloping ground. (Kuhn)

Plate 14.5   A drum mower folded for transport. (Kuhn)

Plate 14.7   This rotary disc mower with seven discs cuts a 3 m swath. The cones on the disc at each side of the cutter bar deflect the cut grass away from the standing crop. (Claas)

DISC MOWERS

The cutting discs on a disc mower are mounted on a bar which runs along the ground. They are gear-driven and timed to prevent the blades on adjacent discs coming into contact with each other. Skids at each end of the bar prevent excessive wear on the underside of the gear housing. The discs run at about 3,000 rpm and each pair of discs contra-rotates; so, for example, a four-disc mower in effect leaves two separate swaths which are merged together to leave a single swath. The discs at each end of the bar have a steel cone that acts as a deflector to keep the cut material away from the standing crop.

Plate 14.8   A sectional view of a gear drive mechanism on a disc mower cutter bar. (Kuhn)

Plate 14.9   This disc mower cutter bar has a spring tine rotor to condition the freshly cut swath. (Claas)

Wider models of trailed disc mower have a centre and two-wing units, which are hinged to the centre section, allowing the wing units to follow uneven ground contours. A tractor with at least 45 kW (60 hp) is required for a 2.6 m cut disc mower with six discs, and a minimum of 60 kW (80 hp) is required for a 3 m cut mounted mower conditioner.

Drum and disc rotary mowers have protective shields on both sides, above and behind the cutting mechanism, to protect bystanders from flying debris. Never walk behind a rotary mower when it is cutting, or allow others to do so.

Some disc mower cutter bars have a heavy balance spring which, when correctly adjusted, allows the cutter bar to follow unlevel surfaces and maintain the correct cutting height. Others have a hydro-pneumatic suspension system which automatically lifts and lowers the cutter bar so that it follows ground contours.

Controls and Adjustments

Always stop the tractor engine before making an adjustment to any mower or swath treatment machinery.

Drive engagement The power take-off shaft lever is used to engage and disengage drive to the cutting discs.

Cutter bar position There are three positions – cutting, turning at the headland and transport – controlled with the hydraulic linkage or an external hydraulic ram. Trailed mowers have a hydraulic ram to lift and lower the cutter bar.

1. *Cutting* The bar is lowered by the hydraulic system or an external ram and the cutting height is controlled with an adjustable skid at each end of the cutter bar. Topping skids, which raise the cutting height to about 100 mm for topping pastures, can be fitted to some rotary mowers. An adjustable balance spring helps the cutter bar follow uneven field surfaces.

2. *Turning* at *the headland* The cutter bar is raised hydraulically at the headland with the outer end lifted higher than the inner end to prevent it picking up grass from the previously cut swath.

3. *Transport* The cutter bar on most mounted mowers is raised to a vertical position for transport with a hydraulic ram. Some mounted models are swung round behind the tractor with a ram. On some trailed mowers, a ram is used to swing the cutter bar behind the tractor and the mower is towed endwise. Some mowers with a cutting width of 3 m or less may be realigned from their offset working position with a hydraulic ram to travel straight behind the tractor.

Pitch The top link is used to adjust the pitch, or angle, of the cutter bar on a mounted mower. It will normally be set to run level but, by shortening the top link, the cutting blades are angled down nearer to the ground to give a very close cut. This setting should not be used on stony fields. The cutting height of a trailed mower is usually altered with a ram at each side of the mower, which is operated from a hydraulic spool valve on the tractor.

Maintenance

The cutting blades, which are either bolted to the discs or drums or secured with quick release clips, must be kept sharp and in good condition. Some mowers have reversible blades. Drive belt tension must be checked at regular intervals. Most belt-driven mowers have three or four vee-belts, side by side on multiple vee-belt pulleys. The belts must always be replaced in sets. When a broken belt is replaced and correctly tensioned, the remaining belts will have become stretched with use and will not transmit full power.

Gearbox oil levels should be checked at regular intervals. Lubricate the mower every day during the season. Watch for oil leaks, especially from the gear case on bottom-driven mowers, as the casing runs close to the ground and may suffer damage in rough conditions.

Safety Devices

Break-away device A spring-loaded mechanism allows the whole cutter bar to swing backwards and upwards if it hits an obstruction. Reverse the tractor to return the cutter bar to the working position.

Slip clutch Some rotary mowers have a slip clutch in the drive from the power take-off shaft to the gearbox. This protects the drive from overload. Vee-belt drives used on some mowers also act as a safety clutch, as the belts will slip if the cutter bar is overloaded.

FLAIL MOWERS

Flail mowers have a high-speed horizontal rotor, with swinging flails which cut the grass and leave it in a fluffy swath. The power take-off driven rotor is enclosed in a metal shield that helps to form the swath. The crop is bruised by the cutting flails; this speeds up the haymaking process but may result in loss of leaf. The drive from the power take-off is transmitted through a gearbox and then by belts or a chain to the flail rotor.

Some flail mowers are mounted behind the tractor; others are offset to one side or front-mounted to keep the tractor wheels off the

Plate 14.10   A front-mounted flail mower for topping pastures and set-aside mowing. A roller behind the flail rotor stops the flails hitting the ground when working on uneven surfaces. (Teagle)

uncut crop. 'Y'-shaped blades, suspended on short chains, are used to cut grass; chisel-edged blades are used for mulching and cutting heavy scrub.

Cutting height is adjusted with skids at each side of the rotor housing. Offset flail mowers are transported with the cutting unit folded back behind the tractor.

MOWER CONDITIONERS

Freshly cut grass has a moisture content of about 75 per cent. For safe storage as hay in bales, this must be reduced to approximately 20 per cent. It is desirable, with the aid of the sun and the wind, to reduce the moisture content as quickly as possible. Hay treatment machinery speeds up the drying process by turning and loosening the swaths. This work should be done with the aim of minimising loss of the valuable leafy material.

Mower conditioners combine cutting the crop with the first treatment in the drying process by bruising the cut grass as it passes through the machine. Some mower conditioners have free swinging nylon fingers on the rotor shaft, which treat the whole crop as it leaves the

Plate 14.11   This trailed rotary drum mower and conditioner cuts and conditions a 3.1 m wide swath. A hydraulic ram is used to swing the mower behind the tractor for transport. The swath between the tractor wheels was cut by 3 m mower conditioner mounted on the tractor's front linkage. (John Deere)

Plate 14.12   The front- and rear-mounted mower conditioners on this tractor have a combined cutting width of 8 m. Hydraulic rams fold the rear-mounted units to a vertical position for transport. (John Deere)

Figure 14.2   The path of grass through a mower conditioner. The crop is passed from the cutting discs to a tined conditioning rotor which bruises the crop against the adjustable hood to hasten drying. (John Deere)

cutting drums or discs, while others have a pair of dimpled rollers. Some conditioners use vee-shaped spring tines on a horizontal rotor which bruise the grass stems to help release the sap as the crop leaves the cutting mechanism. A rotor speed of about 900 rpm is common on many machines, while others have a two-speed rotor which can be set to run at approximate speeds of 600 or 900 rpm.

Mower conditioners, usually with a cutting width of about 3 m, may either be rear-mounted or towed from the tractor drawbar; and both can

be used in conjunction with a front-mounted unit for extra output. Rear-mounted and trailed models are offset and are usually swung round behind or into line with the tractor for transport. High work rates can be achieved with a front-mounted mower conditioner and two rear-mounted models offset to each side of the tractor.

A typical rear-mounted mower conditioner has side deflector plates, which are used to vary the height and width of the swath, typically between 1.3 and 2.4 m. A wide swath promotes rapid drying, but it may be necessary to reduce its width with a side rake before the crop is collected with a forage harvester or a baler for silage.

Trailed mower conditioners are attached to the tractor in a central hitch arrangement which is used to offset the machine with the cutter bar and conditioner when cutting and bring it into line behind the tractor for transport.

Some trailed mower conditioners have deflector plates used to group the swaths cut on consecutive runs close together to make larger swaths for tedding with a wide tedder or picking up with a high output forage harvester.

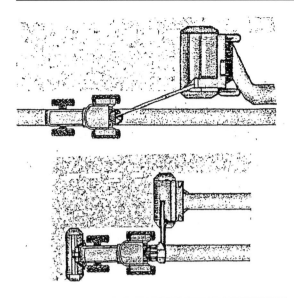

Figure 14.3   Cutting grass for silage using a rotary mower. (Top:) A second swath is placed alongside one cut by a mower on the previous run. (Bottom:) Cutting 3 m swaths with a front- and an offset rear-mounted mower conditioner. (John Deere)

SWATH TREATMENT MACHINERY

Various treatments can be made to the cut swath to improve circulation of air and reduce its moisture content to a safe level for baling.

- *Tedding* The swath is lifted and loosened so that air can circulate freely through the grass.
- *Swath turning* This process involves turning individual swaths and moving them sideways on to fresh ground.
- *Side raking* Two or more swaths, or windrows, are put together to form a much larger one for baling.
- *Spreading* The swath is spread over a wider area to reduce the depth of material and hasten drying.

Some swath treatment machines can be used for most or all of these operations, but others, designed to be used at high-speed, are single- or dual-purpose machines.

Rotary Tedders

There are several types and working widths of mounted and trailed rotary tedders with two,

four or six tined rotors. Wider tedders, with working widths of up to 17.5 m, may have fifteen or more rotors.

Plate 14.13   A hydraulic ram is used to lower the two transport wheels of this eight-rotor rotary tedder. (Twose)

Plate 14.14   A slip clutch protects the power take-off drive on this six-rotor tedder folded for transport. (Twose)

Plate 14.15   Side raking – or rowing up – a hay crop with rotary tedder. (Kuhn)

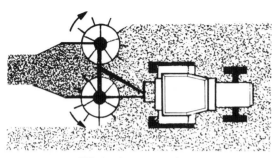

Windrowing a spread crop

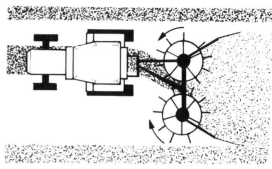

Spreading a swath

Note the position of the rear doors for each operation

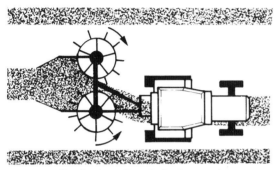

Turning a swath on to fresh ground

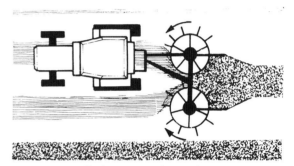

Putting two swaths together

Figure 14.4   Rotary tedder settings for windrowing, spreading and swath turning. Note the position of the rear deflectors for the different swath treatments. (Lely)

A typical rotary tedder (Plate 14.13) with a working width of 7.85 m has eight independently mounted rotors which individually follow undulating ground. The hydraulically adjusted wheels, with low ground pressure tyres, control the height of the double tines, which can be adjusted to vary the distance the tines throw the swath. Hydraulic rams are used to fold the wing sections down to a width of under 3 m for transport.

The twin-rotor rake (Plate 14.14) can be used for spreading, tedding or side raking. Fully floating tines are attached to the lower rim of contra-rotating rotor cages, which are carried on castor wheels. An adjustment on the castor wheels is used to alter the height of the tines, and the adjustable deflector doors provide different widths of swath. Both rotors are set to rotate outwards for spreading and tedding, and when windrowing, or side raking, the rotors are set to turn inwards to form a swath ready for baling.

Rotary Rakes
Gathering spread material into a windrow for baling can also be done with a power take-off driven rotary rake with one, two or four slow-speed rotors with heavy spring steel tines. A cam track on the tine rotor(s) controls the working angle of the tines so that they are almost vertical while raking the crop sideways and then lifted to leave the crop in a neat windrow. The tine arms can be adjusted to suit the volume of crop and leave the required size and shape of swath for the baler.

Plate 14.16   Rotary rakes can be used to gather hay crops into windrows for baling. (Twose)

Plate 14.17   A finger wheel swath turner can be used for high-speed swath treatment. (Lely)

A typical single rotor machine can, for example, be adjusted to rake two 2.7 m wide swaths into a single 1.7 m wide swath at one side of the machine. A typical twin-rotor model with a 7.3 m working width, which will rake three 3 m wide swaths into a single swath 2.15 m wide, is hydraulically folded down to 2.7 m for transport.

Overshot Tedders

Three-point linkage mounted overshot tedders have a horizontal power take-off driven rotor with rows of rotary spring tines under a hood with adjustable rear deflector doors. The tines lift the swath over the tine rotor and return the aerated crop to the ground in a loose swath. The shape and size of the swath will depend on the position of the rear doors.

Finger Wheel Swath Turners

Although they have lost favour with many farmers, finger wheel swath turners with spring-tined walking stick-shaped spring tines on rake wheels driven by ground contact are still used in some areas. Either trailed or mounted on the three-point linkage, they can have between four and sixteen rake wheels, which can be arranged in various ways for spreading, swath turning and side-delivery raking. With working widths of up to 10 m, finger wheel machines can be used for high-speed swath treatment, but they have a tendency to roll the crop into a tight swath.

Care of Hay Making Machinery

Lubrication of the power drive bearings and gearboxes must not be neglected, and tyre pressures should be checked at the start of the season. The tines must be kept tight and any broken tines should be replaced.

CHAPTER 15

Silage Making Machinery

Machinery used for making silage includes disc or drum rotary mowers or mower conditioners, self-propelled or trailed precision chop forage harvesters, trailed double-chop forage harvesters, precision chop forage wagons, round or big square balers, high-sided trailers and a variety of loading equipment. Some livestock farmers prefer to make their own silage, but with the high cost of the capital equipment involved, and the need to do the job quickly to prevent spoilage, many farmers employ a contractor to do the work for them.

The crop is usually direct-harvested, but on some farms it is cut and left in swaths to wilt for a day or so to increase the proportion of dry matter to water and reduce the likelihood of large quantities of effluent leaking from the bales or the clamp. Trailed or self-propelled forage harvesters chop the crop into short lengths and blow it into a trailer, either pulled alongside or towed behind the harvester. The same machines can be used with a special header to harvest maize, which is also chopped and stored in clamps or pits.

Grass for making silage can also be picked up with a forage wagon, which is a self-loading and unloading trailer with a chopping unit just behind the pick-up attachment. Some farmers prefer to bale their grass when silage making with a round or big square baler, which may or may not incorporate a chopping unit. These bales are then wrapped in plastic to exclude air during the fermentation process.

Plate 15.1 A self-propelled forage harvester collecting a swath of wilted grass for silage. After being chopped into short lengths, the crop is blown into a high-sided trailer. (Claas)

SELF-PROPELLED AND TRAILED FORAGE HARVESTERS

Self-propelled or trailed precision chop forage harvesters with a pick-up cylinder, disc mower cutter bar, or a maize header are used for high output silage making operations on many farms. Most self-propelled forage harvesters, with a daily output of 120 ha (300 acres) or more, have four-wheel drive, hydrostatic transmission and, depending on size, an engine in the 200–650 kW (260–880 hp) bracket.

The complete forage harvesting operation is controlled with a joystick and bank of push button switches in the cab. On a self-propelled machine, joystick controls usually include forward and reverse speeds, a feed roll reverser (Figure 15.1) to clear blockages, controls to

Plate 15.2   Harvesting maize with a self-propelled forage harvester. Grass pick-up and maize chopping heads are interchangeable. (John Deere)

Plate 15.3   An offset trailed forage harvester with a 2.1 m wide pick-up filling a high-sided trailer towed behind the machine.

raise and lower the pick-up or cutter bar, rotate the delivery spout and vary the angle of the chute extension. An emergency stop button on the joystick instantly cuts off the drive to the front section of the harvester if there is a blockage or some other problem. Switches are used to control almost every other function, from changing the gear ratio and engaging the main harvester drive to activating the knife-sharpening mechanism. A panel or computer screen terminal monitors various functions including engine and harvesting operations, metal detector, forward speed and fuel level.

Self-propelled forage harvesters can be used with different headers to enable them to harvest other forage crops for silage. For pre-cut grass, a tined pick-up cylinder with working widths ranging from about 2.5 to 4.5 m is used. A rotary disc header or a reel and reciprocating knife header, like that in a combine harvester, can be used to direct-cut standing grass or green cereal crops for silage. Maize grown for silage can be harvested with a maize header that may cut between 4 and 12 rows in one pass.

Trailed forage harvesters with a pick-up attachment varying in width from 1.5 to 3 m and an offset drawbar are still used on some farms. They may have either a flywheel or a cylinder

chopper for chopping the previously cut and wilted crop. Double-chop forage harvesters, which cut and then chop the grass in one operation, are used by some farmers, but as they do not allow for wilting, are not often used for making silage. They are, however, sometimes used for zero-grazing systems where freshly cut grass is taken directly to housed cows.

Precision Chop Forage Harvesters

Precision chop forage harvesters have a chopping cylinder, or a flywheel chopping mechanism, with a set of knives driven by multiple

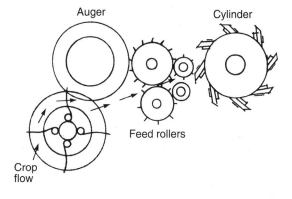

Figure 15.1   Crop flow through a precision chop forage harvester. The feed rolls can be reversed to clear blockages. (Claas)

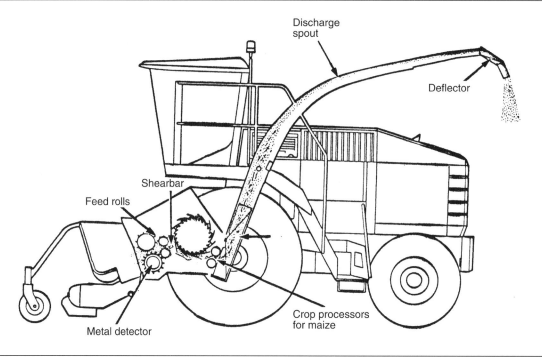

Figure 15.2   The layout of a self-propelled precision chop forage harvester.

vee-belts which chop the crop into precise lengths. The previously cut swath, which is usually left to wilt, is collected by a pick-up attachment and passed between the feed rolls to the chopper unit. The feed rolls compress the crop into a firm wad and pass it to the chopper cylinder or to the flywheel chopper, where it is sliced cleanly against a fixed shear bar. The number of knives on the cutter head and the speed of the feed rolls determine the length of chop.

A blast of air, created by the chopper cylinder or flywheel, blows the chopped material up the delivery spout into a trailer. High output harvesters usually have a second fan to aid this process by providing an additional airflow to help carry large volumes of chopped material into the trailer.

The chopper cylinder on a typical high output self-propelled precision chop forage harvester is 750 mm (30 in.) wide, 650 mm (26 in.)

Plate 15.4   Crop flow through a precision chop forage harvester. The feed rolls (1) pass the crop to the segmented knife chopper cylinder (2). The kernel or cob processor (3) deals with the cobs when harvesting maize, before the blower delivers the crop into the delivery spout (4), which carries it into a trailer. (John Deere)

in diameter and runs at 1,200 rpm. One example has 24 chopper knives mounted in a chevron pattern across the rotor, which chops the crop against a shear bar into lengths varying from 4 to 17 mm. Another model has interchangeable chopper rotors with 40, 48 or 56 knives and a fixed shear bar that chop the crop into lengths varying from 5 to 25 mm.

A typical trailed precision chop forage harvester with a 1.7 m wide pick-up has a chopper cylinder 610 mm in diameter and 463 mm wide running at 975 rpm. It can be used with five or ten rows of knives with three cutters in each row. Ten rows with thirty knives give a chop length of between 6 and 16 mm, depending on the speed of the feed rolls, and with alternate rows of knives removed, the crop will be chopped into lengths varying from 12 to 32 mm. It needs about 75 kW (100 hp) at the power take-off; a larger machine with 590 mm wide chopper cylinder needs up to 130 kW (180 hp) with the power take-off shaft running at 1,000 rpm.

A kernel or cob processor (Plate 15.5), optional equipment for precision chop harvesters, can be used when harvesting maize and

Plate 15.5   A 10-row maize header on this self-propelled forage harvester has rotary dividers to separate the rows being cut from those left standing. The harvester can be automatically steered along the rows. (John Deere)

whole-crop cereals with a high dry matter content. It is used to crack the kernels, which aids digestion of the silage when it is fed to livestock. The crop is passed between two rollers with toothed profiles which gently crack open every kernel of maize. The spacing between the two rollers is either adjusted manually or remotely from the cab.

Some self-propelled harvesters have an automatic steering system with sensors on the header linked to the rear axle steering mechanism that keeps the harvester on the rows and enables the driver to concentrate on harvesting the crop.

Most forage harvesters have a metal-detecting magnetic field built into the lower feed roll which is made with non-magnetic alloy steel. When a piece of ferrous metal (iron or steel) is detected in the crop, an electronic sensor stops the feed roll to prevent damage to the chopper knives, and at the same time activates an audible warning in the cab.

Controls and Adjustments

Pick-up height This is adjusted either with a hydraulic ram or wheels on the pick-up cylinder. Some forage harvesters have both methods of adjustment.

Delivery spout The angle of the spout and direction of delivery can be set for rear delivery or from either side of the machine. The deflector plate at the end of the spout is used to control the discharge angle of the harvested crop as it is blown into the trailer. These adjustments may be made either hydraulically or with a small electric motor. Older models usually have a manual adjustment.

Feed roll A reverse drive is provided on most forage harvesters. This enables the driver to clear blockages, the main cause of which is overloading the machine.

Shear bar Adjustment is provided to move the shear bar close to the chopper knives to achieve maximum cutting efficiency. The shear bar setting is either adjusted manually or controlled remotely from the cab.

Length of chop This is normally between 5 and 25 mm, but can be longer. Short grass, which clamps more easily, aids the exclusion of air from the fermentation process, but the animals will digest the silage too quickly if the grass is too short. Chop length depends on the number of knives on the chopping cylinder or flywheel and the speed of the feed rolls. Some forage harvesters have an infinitely variable drive system for changing the speed of the feed rolls; others have a manual gearbox.

Double-Chop Forage Harvesters

Trailed either behind a tractor, or from an offset drawbar, double-chop harvesters have been replaced by precision chop forage harvesters on many farms, but are sometimes used on farms where zero-grazing is practised and fresh cut grass is delivered directly to the cows. Double-chop harvesters may either have a pick-up attachment for collecting previously cut and wilted crops or a flail cutting mechanism for direct harvesting. A cross-conveying auger feeds the grass collected by the pick-up or the cutting mechanism to a flywheel chopper with a set of knives and paddle fan blades. Depending on the model and the required length of chop, the flywheel chopper runs at speeds of 700–1500 rpm. Flywheel chopper forage harvesters have feed rolls which compress the material to give an even feed to the chopping unit. The chopper knives cut the crop against a stationary knife or shear bar in the opening from the auger to the flywheel housing. After chopping, the fan blades blow the crop up the delivery spout into a trailer.

There is some variation in the range of chop lengths available with different harvesters. One double-chop forage harvester cuts the crop into lengths varying from 5–22 mm; another has chop lengths ranging from 14–35 mm.

Controls and Adjustments

Length of chop This is varied by altering the speed of the chopper flywheel, changing the number of chopper knives or by varying the speed of the feed rolls. A slow chopper cylinder speed with a full set of knives gives the shortest chop length.

Pick-up or cutting height This is altered with a hydraulic ram or with adjustable wheels on the pick-up cylinder. Some trailed machines have a height control ram on the drawbar.

Feed roll Blockages in the feed rolls, mainly caused by overload, can be cleared by reversing the drive.

Delivery spout The spout can be positioned to blow the chopped material into a trailer pulled alongside or behind the harvester. The deflector plate on the end of the delivery spout can be adjusted to control the discharge angle and blow the crop into different parts of the trailer.

Drawbar This is used to offset the harvester to one side of the tractor when in the field and position the harvester behind the tractor for road transport.

Safety Devices

All forage harvesters have safety mechanisms to protect them from mechanical damage. They include:

- *Main drive protection* is provided by a slip clutch and an over-run clutch. The slip clutch allows the tractor end of the power take-off shaft drive to continue turning if the harvester stops through overloading or there is a blockage in the machine. The over-run clutch allows the machine to slow down at its own pace when the power drive is disengaged.
- *Metal detector* The chopper knives and shear bar may be damaged if a forage harvester picks up a piece of ferrous metal concealed in the swath. When the magnetic metal detector in the feed rolls locates a piece of iron or steel, it instantly stops the entire feed mechanism.

The metal object can be removed by engaging reverse drive while the machine is stationary.

- *Vee-belt drive* A vee-belt drive provides an automatic built-in safety mechanism, because the belt will slip when the drive is overloaded.

Attachments

Grass headers Most forage harvesters have a tined pick-up cylinder to collect and feed previously cut and wilted grass crops into the chopping cylinder or flywheel. Attachments for harvesting maize and whole-crop cereals can also be used on many forage harvester.

Maize headers One type of maize header has dividers which guide individual rows of the crop to rotary cutters, where the stems are cut and the crop is passed by an auger to the feed rolls. Another type, with smaller crop guides, can be used to cut along or across the rows. The guides direct the crop to large rotary cutters, where the stems are severed and carried to the feed rolls. A kernel, or cob processor (Plate 15.4), is used to crack the individual grain kernels to make maize silage more digestible when fed to cattle. Maize headers are made for harvesting 4, 6, 8, 10 or 12 rows at a time. Some maize headers have an automatic row-following steering system.

Whole-crop headers These may either be similar to a combine harvester cutter bar or a disc mower cutter bar. They are used on some farms to harvest green cereal crops for silage. Some models have a hydraulic motor-driven pick-up reel and an auger with a reversing mechanism for clearing heavy blockages on the cutter bar table.

Maintenance of Forage Harvesters

Lubrication points must be serviced daily during the harvest season. Follow the lubrication schedule given in the operator's handbook. Many self-propelled forage harvesters have central lubrication banks with groups of conveniently placed grease points connected to the more inaccessible bearings.

Knife sharpening Most precision chop and flywheel chop forage harvesters have a built-in knife sharpener. Some machines have a knife-sharpening unit which is controlled automatically from the driving cab. A grinding stone running on guide rails, which are parallel to the edge of the knives, is drawn across the face of the chopper cylinder either by a hydraulic ram or by hand. The chopper blades are turned at slow speed, either under power or by hand, while the stone is drawn across the knives.

Plate 15.6   The chopper cylinder, which runs very close to the shear bar at 1,200 rpm, chops the harvested crop into short lengths. (Claas)

Plate 15.7   Sharpening the chopper cylinder knives with an automatic sharpener. A grindstone moves along a rail above the cylinder, which is rotated at a slow speed in reverse to give the knives a keen cutting edge. (Claas)

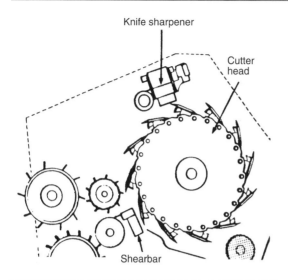

Figure 15.3   The knife-sharpening mechanism on a precision chop forage harvester.

Plate 15.8   Forage wagons pick up, chop and load swaths of previously cut grass in a single operation. A chain and slat floor conveyor is used to discharge the load at the silage pit or clamp. (Claas)

The shear bar must also be sharpened and adjusted from time to time to ensure a clean cut. The clearance between the shear bar and the knives is critical. This is usually adjusted manually, but some self-propelled harvesters have a small electric motor, controlled from the cab, to adjust shear bar clearance.

The cutting flails, which cut and chop the crop, must be secure and sharp. Any missing or badly damaged flails must be replaced to avoid rotor vibration. Flails should be replaced in opposite pairs to maintain correct rotor balance.

FORAGE WAGONS

A forage wagon, which is a self-loading and unloading trailer with a chopping unit, is an alternative to a trailed or self-propelled forage harvester. Silage making with a forage wagon requires much less labour and machinery, as the harvesting operation is self-contained and no other tractors or trailers are required except on the clamp.

A pick-up attachment at the front of a high-sided wagon lifts the crop to feeder tines and a chopping mechanism. Some wagons have an automatic knife-sharpening mechanism controlled from the tractor cab. The chopped material is passed into the wagon, and when fully loaded it is hauled to the silage clamp or pit. After opening the rear door of the wagon, usually with a hydraulic ram, the load is discharged by a moving floor conveyor. Some forage wagons have a cross-conveyor that can be used for direct feeding into cattle troughs.

Forage wagon capacities typically range from 10 to 70 cubic metres. For example, a 30 cubic metre wagon will hold about 20 tonnes of grass and, depending on the number and setting of the knives, give chop lengths of between 40 and 75 mm. The cutting unit can be withdrawn on some wagons so that it can be used to pick up and load swaths of dry grass.

BUCKRAKES

Until recent times, a buckrake, mounted on the rear three-point linkage, was used on almost every farm where silage was made. While some farmers still use a buckrake to fill silage clamps, the increased use of high output self-propelled forage harvesters and wrapped silage bales has restricted their use. The economic use of a tractor-mounted buckrake is limited to larger units now used behind or on the front of high-hp tractors.

Plate 15.9   This buckrake has a back board, operated with a hydraulic ram, which is used to push the load of grass or other green material off the tines when filling a pit or clamp. (Johnston Brothers)

They have been almost entirely replaced with large forks fitted to telescopic handlers and wheeled loaders, but large buckrakes have come back into favour for use on the front linkage on more powerful tractors.

A buckrake consists of a number of strong steel tines, about 1.4 m long, bolted to a tine bar. Some buckrakes pivot on a frame attached to the three-point linkage, but more modern examples are of the push-off design with a hydraulic ram and a pusher gate.

A buckrake is used by reversing it into a heap of grass tipped from a trailer. The tines are pitched (angled) slightly downwards to give a clean pick-up. Too much pitch may result in damage to the concrete floor of the silage clamp. The tines are tipped or the push-off ram is used to drop the load where required on the silage clamp or pit.

SILAGE HANDLING AND STORAGE

Precision-chopped silage made by using a forage harvester or wagon is usually 'clamped' in a well-compressed heap on a concrete pad, before being covered with plastic sheeting to prevent the ingress of air which will spoil the silage. Alternatively, it can be baled using a high-density round

or big square baler, then wrapped in plastic film, to achieve the same result.

Pits and Clamps

Telescopic fork lifts, wheeled loaders, tractor front-end loaders and tractor or telehandler-mounted buckrakes can be used to fill silage pits or build clamps. Loading and consolidating silage with any machine, especially when making an above-ground clamp, presents certain dangers. Never drive close to the edge when consolidating the clamp. Wide wheel settings will increase the stability of the machine.

Bale Wrappers

Wrapping has largely replaced bagging as the preferred method most farmers use to preserve and protect baled silage and exclude air from the ensiled material. A bale wrapper may be an independent machine or form part of a baler/wrapper combination. With independent wrappers, wrapping with plastic film can be done in the field or after the bales have been transported to the farmyard. When they are wrapped in the field, the bales can be left there until it is convenient to haul them to the farmyard. This may be done with a tractor loader or a telescopic forklift equipped with a squeeze grab or similar attachment, which loads and transports the bales without damaging the plastic.

Plate 15.10   Loading wrapped silage bales with a squeeze grab attachment on a tractor loader. (John Deere)

CHAPTER 16

Balers

Balers pick up loose hay or straw, or wilted grass for silage making, and compress it into bales of an even size and weight. The completed bale is tied with twine, or often with net or plastic film if a round baler is being used. The completed bale is discharged from the back of the machine where it either falls to the ground or, in the case of large or small square bales, it may pass into a bale sledge or collector towed behind the machine. When a sledge or collector is used, groups of bales are left at intervals across the field. There are many different systems for handling and carting bales to the store.

There are three types of pick-up baler. Most farmers now use either a big round or a big square baler or employ a contractor to do this work for them. Bale weight will depend on its size and the type of material being baled. Big round straw bales weigh between 240 and 400 kg and hay bales range from 350 to 500 kg. Because of its high moisture content, a 1.5 m diameter silage bale will weigh between 500 and 900 kg or more. Big square straw bales may weigh between 120 and 600 kg, and hay or silage bales between 200 and 900 kg. The third type of pick-up baler makes small rectangular bales weighing about 25 kg when baling straw and 35 kg when making hay bales. This type of baler is not used to make bales for silage.

An important advantage of a big round or big square baler is the low number of bales made per hectare. This makes handling from field to store a simple task, provided that suitable mechanical handling equipment is available. Straw merchants prefer big square bales

Plate 16.1 Big square balers are made in various sizes. The larger models make bales 1.2 m wide, 1.3 m high and up to 3 m long. The bales, tied with heavy-duty twine, can weigh over a tonne when working in silage. (Claas)

as a lorry can be loaded very quickly without any wasted space.

BIG ROUND BALERS

Big round bales vary in size and weight depending on the type of baler and the crop being baled. Round balers may have a fixed or variable diameter bale chamber. Most round balers with a fixed diameter bale chamber make bales 1.25 or 1.5 m in diameter and 1.2 m wide. Round bales made in a variable-size chamber baler are usually between 1.2 and 1.7 m wide and, depending on model, bale diameter can be varied within a range of 0.9 and 2 m. Because of their heavier weight, hay and silage bales usually have a maximum diameter of 1.2 m.

Plate 16.2   The bale diameter can be varied when using a big round baler with an endless belt bale chamber. (Kuhn)

Plate 16.3   The pick-up cylinder and chopper rotor on a round baler gathers the crop, chops it into short lengths and feeds it into the bale chamber. (John Deere)

The pick-up attachment may, depending on the model, have a pick-up width of between 1.2 m and 2.3 m wide. The tractor and baler wheels straddle the swath and the pick-up lifts the crop either to a rotary chopping system or to feed rolls which carry it into the bale chamber. Depending on the width of the bale chamber, the chopper rotor, if fitted, will have between 12 and 35 knives arranged in four or more spiral rows across the full width of the bale chamber. These knives chop the crop against a row of fixed knives before it passes into the bale chamber. The number of knives used on the chopper will depend on the required length of chop. Used mainly when baling green material for silage, the chopper unit cuts the crop into short lengths, typically between 40 and 70 mm. Chopped grass makes a denser, and therefore heavier, silage bale. The chopper knives are folded away hydraulically on some machines; on others they are removed from the rotor. The chopper blades are individually spring-loaded to protect them from large stones or blockages.

The power take-off drive on a round baler is protected from damage by an overload clutch, while shear bolts protect the pick-up and main drive. There is a relief valve in the hydraulic bale chamber tensioning mechanism on variable diameter balers.

Plate 16.4   Forming the bale in an endless belt variable diameter bale chamber. (John Deere)

Variable Bale Chamber Round Balers

The bales are formed by a number of heavy-duty endless rubber belts carried on rollers in the bale chamber. The formation of a new bale starts when the pick-up attachment feeds the swath into the empty bale chamber. The belts curl the new intake of grass or straw to form the initial core of the bale (Figure 16.1 – Diagram A). As more of the crop is fed into the baler, the core rotates against the bale-forming belts, and the bale grows in size. The centre, or core, is usually less dense than the rest of the

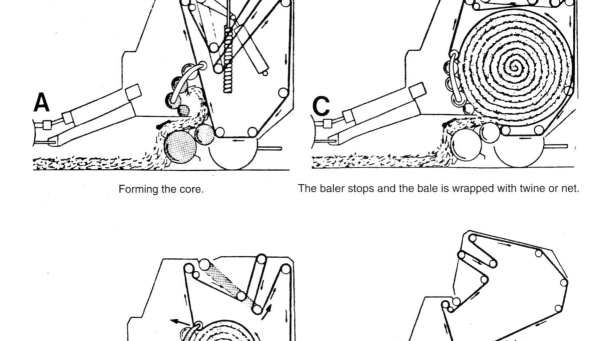

A   Forming the core.

C   The baler stops and the bale is wrapped with twine or net.

B   The bale chamber expands as bale size increases.

D   The tailgate opens and the bale is ejected.

Figure 16.1   Forming a bale in an endless belt variable diameter bale chamber.

bale. This is desirable for hay bales as it allows air to pass through the hay and keep it sweet in storage. There is a risk of the hay becoming musty if the bale has a tight core. For straw bales, which are often stored outside without protection, the dense outer layers help to prevent water ingress and aid run-off.

The crop is rolled round and round by the belts (Figure 16.1 – Diagram B). As more material is fed into the baler, the tension arm carrying two of the belt rollers moves upwards on its pivot to relax some of the pressure on the belts, which allows the bale chamber to

expand to accommodate the increasing bale diameter. Variable bale chamber round balers have an adjustable actuator which controls the bale-forming belts and limits the bale diameter, with two hydraulic rams being used to control the size of the bale chamber by tensioning the belts (Figure 16.2). As the diameter of the bale increases, it puts more tension on the belts, and this in turn increases the oil pressure in the hydraulic density control system. When the pressure reaches a pre-set level, the relief valve releases some of the pressure in the ram and the belt tensioning arms allow the bale chamber

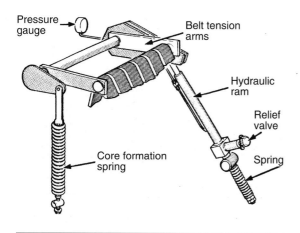

Figure 16.2   Hydraulic bale chamber tensioning system. (New Holland)

Plate 16.5   Fixed diameter bales are made by a big round baler with a roller type bale chamber. (John Deere)

to expand. An adjustable spring (Figure 16.2) controls the density of the core, and when this is made, the hydraulic density control comes into operation.

An adjustable relief valve in the hydraulic density control unit is used to vary bale density. Some round balers have an in-cab computer that can be set to give the optimum bale density for different types of crop including hay, straw and silage.

An indicator on the baler warns the driver when the bale reaches its pre-set diameter (Figure 16.1 – Diagram C). Most round balers have to stop while the bale is wrapped with several turns of heavy twine or wrapped in net because it is not possible to feed material into the bale chamber during the bale wrapping process. Another design of round baler has a pre-compression or holding chamber for the crop and does not have to stop during the wrapping process.

With the wrapping complete, a hydraulic ram opens the tailgate (Figure 16.1 – Diagram D), the bale rolls down a spring-loaded ramp to the ground, the rear door is closed, the tractor moves off and starts the next bale.

Another type of variable bale chamber has an endless slatted chain conveyor with closely spaced slats attached to a drive chain at each side of the bale chamber. As the bale size

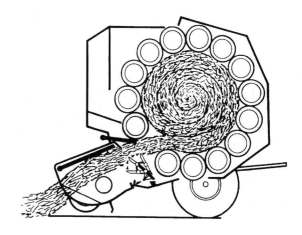

Figure 16.3   Fixed diameter roller type bale chamber. (Claas)

increases, the chain tension arms pivot, allowing the bale chamber to increase in size until it has reached its pre-set diameter. The bale is then secured with twine or net and ejected from the rear tailgate.

Fixed Bale Chamber Round Balers

This type of round baler may have a bale chamber with a set of endless belts, an endless chain and slat conveyor or, more commonly, a series of dimpled steel rollers. Some fixed bale chamber round balers have a chopping rotor which

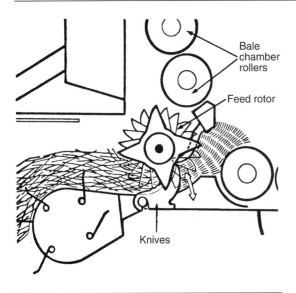

Figure 16.4   Crop flow from the pick-up through a chopper rotor to a fixed diameter roller type bale chamber. (Claas)

cuts the crop into short lengths in order to make tightly compressed bales. Round balers with a fixed bale chamber may make 1.0, 1.2 or 1.5 m diameter bales.

Roller type bale chambers may have between 16 and 22 dimpled steel rollers driven by a heavy-duty chain. As soon as the crop enters the bale chamber, the rolls begin to form a bale, and the rolling process continues until the bale reaches its full diameter, when baling stops while the bale is wrapped with either twine or net and ejected from the rear door.

Endless belt bale chambers may have four, five or six separate heavy-duty belts carried on rollers across the full width of the chamber. The pick-up feeds the swath into the bale chamber, and as the bale grows in size, the crop comes into contact with an increasing area of the belts until it reaches its full diameter. At this point, baling stops and twine or net is wrapped round the bale before it is ejected onto the ground.

The bales in a fixed diameter chain and slat big round baler are formed in a similar way to an endless belt baler, with the endless chain

and slat conveyor running round the inside of the bale chamber. When the bale reaches its full size, baling stops while it is wrapped and ejected from the rear tailgate.

Wrapping the Bales

Round bales can be wrapped with twine or net. When the bale is made, an indicator, either on the baler or in-cab monitor screen, shows that it is ready to be wrapped. With most balers, forward travel must cease while the bale is wrapped, but some round balers have a pre-compression chamber that holds back the crop from the pick-up cylinder so that baling can continue during the wrapping process.

Twine is wrapped spirally around the bale, then cut, and the end is tucked into the bale. Some round balers have a single twine arm which travels across the full width of the bale. Others have a twine arm at each side of the bale chamber, with both of them wrapping twine from the outside to the centre and back again. The driver can select the number of twine wraps. The bale will be more secure if several wraps are used, but this will increase the cost of the twine for each bale. The twine box on most round balers holds between eight and twelve balls of twine.

Net is used as an alternative to hold the bale together before it is ejected from the bale chamber. The net wrapping attachment with a large roll of net automatically wraps the bale when the drive mechanism is engaged. The net is stretched as it leaves the roll to ensure the bale is tightly wrapped. When sufficient net has been wrapped around the bale, it is cut and the bale ejected. Additional rolls of net are carried on the baler.

Silage bales, made from mown and wilted grass or other forage crops, are wrapped with plastic film to exclude air from the ensiled material. This is a necessary procedure in the making of good quality silage. Bale wrappers may be an integral part of big round balers or a separate trailed or tractor-mounted machine

Plate 16.6   A big round baler with an integral bale wrapper. (Claas)

Plate 16.7   A net-wrapped hay bale.

(see page 192). When the wrapper is an integral part of the baler, the completed bales pass from the bale chamber on to the wrapping table, where the bale is rotated and wrapped with about six layers of film in approximately 30 seconds. The wrapped bale is then discharged on to the ground for later collection. When a separate trailed or mounted wrapper is used, the bales may be wrapped either in the field or at the stack.

Controls and Adjustments

Pick-up The tines should be set low enough to gather the entire crop without digging into the ground. A hydraulic ram is used to raise and lower the pick-up into and out of work.

Bale diameter Variable diameter round balers have a hydraulic actuator which controls bale diameter.

Bale density This is altered by varying the tension on the bale chamber-forming mechanism, either with the use of spring pressure or a hydraulic density control system.

Electronic Control Systems

Many big round balers have a monitoring and control system linked to a computer. A control panel in the tractor cab, linked to the baler, tells the driver exactly what is happening in the bale formation and wrapping cycles, the number of bales made, the amount of twine or net wrapped on each bale and how much net or twine has been used. The unit has a memory which, once set, will repeat the same baling cycle to give the required number of wraps of twine, net or plastic film and the bale diameter on variable chamber round balers. The control unit also opens rear door, ejects the bale and closes it again.

BIG SQUARE BALERS

High capacity big square balers have a 2.0–2.5 m wide pick-up and, depending on their size, make bales between 0.8 and 1.2 m wide, 0.5 to 1.3 m high and up to 3.0 m long. Unlike conventional balers, which make small bales and usually have an offset pick-up, big square balers run in-line behind the tractor. Depending on size, big balers with work rates in excess of 40 ha (100 acres) in a day require a tractor of about 70 kW (100 hp) and a power take-off speed of 1,000 rpm.

The weight of a bale depends on its length, cross sectional area and the material being baled. A well-packed straw bale weighs about 160 kg per m³. Big square hay bales weigh between 200 and 270 kg per m³ and green silage

Plate 16.8   Big square balers make straw bales weighing up to 600 kg and over a tonne when baling silage.

Plate 16.9   This big square baler, with a 2.26 m (89 in.) wide pick-up and an automatic bale density control system, makes bales 875 mm (35 in.) high and 1,200 mm (47 in.) wide. (Massey Ferguson)

bales can weigh up to 700 kg per m³. Big square balers make much heavier bales when they are used with a built-in chopper unit behind the pick-up cylinder, which is similar to the chopper on a big round baler (see page 260).

A set of packer tines feed the chopped or full length material under pressure into the bale chamber, where the ram, running at between 30 and 60 strokes per minute depending on model, forms the bale.

Plate 16.10   With a ram speed of about sixty strokes per minute, the knotters on a big square baler have less than a second to tie the knots with heavy-duty twine. (Claas)

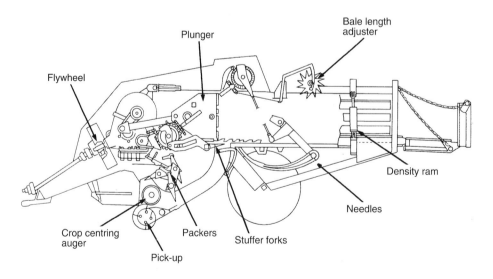

Figure 16.5   Big square baler. (New Holland)

Four, five or six knotters, again depending on model and bale cross-section, are used to tie heavy-duty twine around the bale when it has reached the pre-set length. Many big balers have a fan to keep the knotter deck clear of chaff and dust to ensure efficient knotter operation and avoid the risk of fire.

Crop Flow

The pick-up attachment lifts the crop and short cross-conveying augers move the material to the centre of the pick-up deck. The packers feed the crop into the pre-compression or charge chamber against the holding fingers. When the pre-compression chamber is full and the plunger is fully forward, the stuffer fork drive is engaged and the crop is lifted into the bale chamber, where it is compressed against the crop that is already present. Depending on model, the ram makes between 40 and 65 strokes per minute. When the bale reaches its pre-set length, it is tied and ejected from the bale chamber.

Controls and Adjustments

Pick-up This is raised and lowered by a hydraulic ram. An adjustable spring allows the pick-up to float on uneven ground, and side wheels are used to alter the height of the tines.

Bale density A self-contained system of hydraulic rams in the bale chamber restricts the passage of the material through the machine. Density is varied with either a manual or automatic control unit. The automatic system has an electro-hydraulic valve which maintains the pre-set pressure in the hydraulic ram cylinders. For manual control, the driver uses a valve to vary the hydraulic pressure to suit the swath being baled.

Bale length is adjusted with a stop on the knotter trip arm operated by a star wheel, turned by the bale as it moves along in the bale chamber. The knotter drive may be either engaged mechanically or by an electronic switch.

Plate 16.11   The monitoring screen for this big square baler provides the driver with information about its operation and performance. (John Deere)

In-cab control and monitoring is widely used with big square balers. Bale density can be set and maintained automatically to suit changing crop conditions. Operation of the packers, ram and knotters is monitored on an in-cab screen, and adjustments can be made from the cab. Ram strokes per minute, power take-off speed, pick-up cylinder speed, bale count and other factors can also be checked on the monitor screen.

Using Big Balers

Careful preparation of the swath with a regular shape and even density is important. It may be necessary to rake two or more swaths together when a big baler is used behind a mower or combine harvester with a narrow cutter bar. The tractor wheels must be wide enough to straddle the swath.

Silage bales can be made with most big square balers. After the wilted swath has been baled, each bale, usually no more than 1.2 m long, is wrapped with plastic film to exclude the air (see page 184).

SMALL SQUARE BALERS

Small square balers, also known as conventional pick-up balers, make small rectangular section bales weighing about 25 kg in straw, and are now restricted to specialist use such as baling hay for horses. Typical dimensions for

Plate 16.12 Bales weighing between 12 and 35 kg can be made with a small square bale pick-up baler. (John Deere)

Plate 16.13 The spring-loaded crop guide above the pick-up guides the hay or straw to a cross-conveyor which feeds it into the bale chamber.

bales made with this type of baler are 360 mm high by 460 mm wide and 920 mm long. Bale length can be varied from 0.3–1.4 m, but for easy stacking the bale length is usually twice its width.

Small square balers have a pick-up attachment, like that on other types of baler, which lifts the swath from the ground up to the packer tines or other type of side conveying mechanism. The pick-up is usually offset to the right, but a few balers have the pick-up in-line with the tractor. A spring-loaded crop guide, above the pick-up, holds the crop down against the pick-up tines, and in blustery conditions it also prevents the wind blowing the crop off them.

USING BALERS

The packer tines (Figure 16.6) feed the crop into the bale chamber, where it is compressed by the ram (or plunger) driven by a heavy crankshaft and connecting rod from a gearbox at the front of the baler. Depending on model, the ram makes between 80 and 100 strokes per minute. Each wad (or charge) of material entering the bale chamber is cut by a knife on the pick-up side of the ram. It cuts the crop against a fixed knife at the back of the feed opening into the bale chamber. A star-shaped metering wheel measures the length of the bale being made, and when it reaches the pre-set length, the tying mechanism is tripped (engaged) to start the tying cycle with the needles entering the bale chamber.

There are two knotters. When the previous bale was tied, one end of both twines was retained in the knotters as the needles returned down through the bale chamber to their rest position. As the new bale is formed in the bale chamber, both twines are wrapped around the length of the bale (Figure 16.7). The tying cycle begins when the needles travel up through the bale chamber and place the other end of the twine in the knotters.

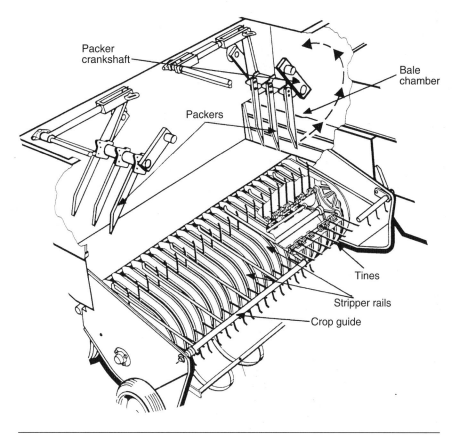

Figure 16.6   The pick-up and packers on a small square pick-up baler. (Claas)

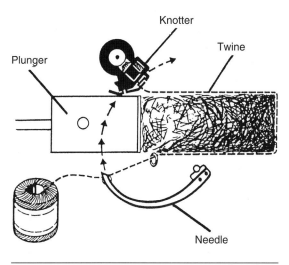

Figure 16.7   Path of the twine to the knotters. (Claas)

Tying the Knot

The tying cycle is completed in about half a second; this is less time than it takes the ram to make one stroke. The main parts of a knotter (Figure 16.8) are the bill hook, retainer disc, stripper arm and the knife. They are gear-driven and timed to operate at fixed points in the tying cycle. When the knotters are tripped, both needles enter the bale chamber and place the twine over the top of the knotter bill hooks and into the retainer discs, which turn slightly to grip the twine. There are now two separate twines wrapped round the bale with both ends of each one held in the retainer discs. The knotter drive gears turn the bill hook, which wraps the twine around itself. The bill hook jaws then open, take hold of the twine and close again to grip the twine.

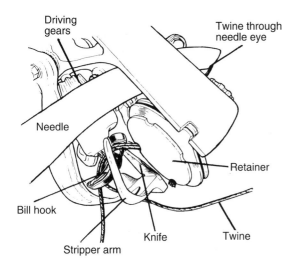

Figure 16.8   A pick-up baler knotter.

The knife is attached to the stripper arm. At this point in the tying cycle, the stripper arm swings across and pulls the twine off the bill hook in such a way that it forms a knot. The twine is cut at the same time, leaving the end of the twine in the retainer disc ready for the next knot. Some knotters have a fixed knife, and the twine is cut as the stripper arm pulls it across the knife.

After tying the knot, the needles withdraw from the knotters and return to their rest position below the bale chamber. The ends of the two twines held in the retainer discs pass down through the bale chamber and needles into the twine box.

CONTROLS AND ADJUSTMENTS

Engaging the drive Pick-up balers are power take-off driven and drive is engaged with the power take-off switch or lever on the tractor.

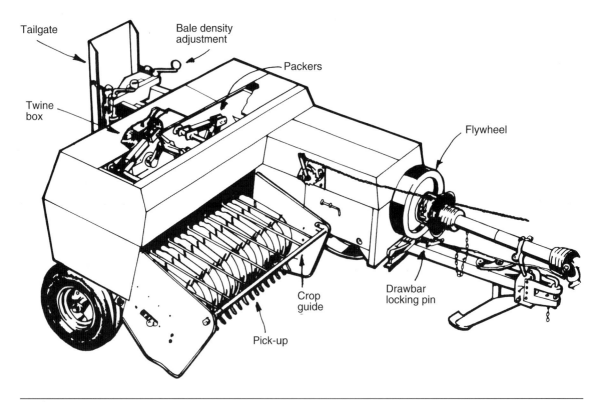

Figure 16.9   The parts of a pick-up baler.

Pick-up height This is usually adjusted hydraulically, or on some older balers, with a hand lever. The tines should be about 100 mm above the ground: low enough for a clean pick-up, but not so low that soil and stones are collected with the swath. An adjustable balance spring helps the pick-up follow uneven field surfaces.

Packer adjustment The packers, which feed the crop into the bale chamber, can be adjusted to suit different crops and swath densities. They should place the crop centrally into the bale chamber to produce a well-shaped bale. If the packers feed too much material to either side of the bale chamber, curved or banana-shaped bales will be made. Adjusting how far the packers feed the crop into the bale chamber will correct this fault.

Bale density This determines the weight of the bale. It is altered with the bale density adjusting screws at the end of the bale chamber, which increase the resistance against the bales as they move along the bale chamber. Both screws must be at the same setting or curved bales may be made. Wedges can also be bolted to the sides of the bale chamber to produce even heavier bales.

Bale length Altered with an adjustable stop on the star-shaped bale length metering wheel.

Drawbar There are two baler drawbar positions. One is used when baling to offset the pick-up so that the tractor wheels run on cleared land. The second is the transport position, with the baler placed in-line behind the tractor.

Line of Drive

The power take-off drive is transmitted through a heavy flywheel to a bevel drive gearbox. A crankshaft on the gearbox output shaft reciprocates the ram in the bale chamber, and a separate shaft from the gearbox operates chain drives to the pick-up cylinder, packers and knotters. The chain drive to the packers is timed with the ram so that the tines feed the crop into the bale chamber as soon as the ram has cleared the feed opening. The tying cycle is also chain driven and is timed to prevent the ram from damaging the needles when they enter the bale chamber.

Pick-up balers have a number of protective mechanisms in the driveline to reduce the risk of mechanical damage caused by a blockage, overloading or mis-timing. A shear bolt, which breaks under a shock load, is located in the flywheel hub to protect the main drive. Some balers have a shear bolt to protect the chain drives to the packers and the knotters. A friction slip clutch on the flywheel absorbs less violent overloading. A flywheel over-run clutch, usually included in the main drive, acts like a bicycle freewheel, allowing the baler to slow down at its own pace after the power take-off drive has been disengaged.

The needles are protected by a ram stop, a heavy steel lug, which prevents the ram damaging them if they enter the bale chamber at the wrong time. The flywheel shear bolt will break when the ram hits the stop. Other baler protective mechanisms include a slip clutch or an over-run clutch on the pick-up and friction brakes which are used to hold the needles and knotter drives in the rest position between tying cycles.

Pick-Up Baler Maintenance

Many baler bearings must be greased daily during the season, but some are sealed units which require no attention during their working life. Gearbox oil levels and tyre pressures should be checked at frequent intervals. Drive chains and belts must be at the correct tension and chain drives should be lightly oiled. The ram, which runs on adjustable runners or rollers in the bale chamber, should be set so that there is a clearance of about 1 mm between the fixed knife and the ram knife. The bales will have a ragged edge if the ram knife is blunt or there is too much clearance between the knives.

Never attempt to clear a blockage in a pick-up baler without first stopping the tractor engine. Disengaging the power take-off drive is not enough, because someone could accidentally re-engage the drive and the consequences could be serious.

Using Pick-Up Balers

Where possible, a cleaner pick-up of the swath will be achieved by baling in the opposite direction to that of the combine or side rake which produced the swath. An even feed into the bale chamber is essential for well-shaped bales. Hay bales are much heavier than straw bales and will need less tension on the bale density screws. When the swath is a little damp, the density adjusters must not be over-tightened or the bales may heat. Tying problems may be due to a knotter fault, but first check that the twine is correctly threaded from the twine box to the needle eyes. Make sure that the twine guides are not restricting the flow of twine.

BALE HANDLING EQUIPMENT

Various machines can be used to handle all sizes of round and square bales in the field. Bales used for silage are wrapped with plastic film by a bale wrapper, which may either be attached to the baler, or a separate machine. Bale collectors range from sledges towed behind small square balers, which collect the bales as they leave the bale chamber, to sophisticated and expensive machines that pick up and automatically load the bales onto a purpose-built carrier or trailer. Accumulators are used to form groups of two or three big square bales on some farms, but round bale accumulators are less common.

Plate 16.14 Completed bales are transferred on to the wrapping table on this round baler/wrapper combination which, using 500 or 700 mm wide plastic film, can wrap up to 50 bales in an hour. When not required, the wrapper can be removed from the baler. (Kuhn)

BALE WRAPPERS

Wrappers are made to wrap round or big square bales. These machines consist of a revolving multiple roller floor, a spool for a roll of plastic film, one or two wrapping arms and a cutting mechanism. Bale wrappers may either be an integral part of a big baler, usually a round baler, or a separate wrapper, which may be either trailed or mounted on a tractor three-point linkage or a telescopic handler. On balers with an integral bale wrapper, the completed bale is guided from the bale chamber on to the wrapping table. A pair of wrapping arms cover the bale with between four and six layers of net or plastic film as it rotates on the wrapper table. Silage bales are normally wrapped with four to six layers of 500 or 750 mm wide plastic film with a 50 per cent overlap between each layer. Twenty-four revolutions of the wrapping table are needed to apply six layers of plastic film. After it has been wrapped, the bale is lowered to the ground. Some integral bale wrappers with a wrapping table mounted immediately behind the baler do not need the baler to stop during the wrapping process.

Plate 16.15 Bale wrappers include those which are designed to be trailed, carried on the three-point linkage or mounted on a telehandler. (Kuhn)

Portable bale wrappers may be driven by a hydraulic motor or the tractor power take-off. Used either in the field or farmyard, they operate in the same way except that the bale has to be lifted on and off the wrapping table. Single big bales are lifted from the ground on to the wrapping table, either with a hydraulically operated loading arm on the machine if it is being towed around the field or with a bale-loading grab on a telescopic fork lift.

Bale Collectors and Loaders

Small square bales can be collected with an accumulator or a sledge towed behind the baler, which either leaves them in a random heap, or more usually in a 'flat eight' with two rows of four bales side by side. The bales can then be loaded on to a trailer with a flat eight grab attachment – a frame with a series of hydraulically operated claws – on a tractor front-end loader or telescopic handler. This type of accumulator operates mechanically with a series of spring-operated gates to form the flat eight pattern.

Another type of automatic accumulator, towed behind a small square baler and powered by the tractor hydraulic system, collects the bales as they leave the baler and stacks

them in groups of 14 to 21, depending on the machine. They are bound together with thick twine or metal bands. Baling continues non-stop while the completed pack of bales is pushed from the accumulator on to the ground and can be moved with a telescopic handler.

Accumulators are also made for big square balers. They carry and dump groups of two or three big bales on the field for later collection, which reduces field traffic movements collecting them.

Plate 16.16 This bale accumulator forms packs of 18 or 21 small square bales and secures the bundle with two half-inch wide metal bands. (Bale Band-It)

Plate 16.17 This mechanical bale accumulator collects the bales as they leave the baler and arranges them in a flat eight formation. (Ritchie)

Plate 16.18   A flat eight bale grab attachment for a tractor front-end loader or a telescopic fork lift is used to pick up and load bales left in a flat eight formation behind a baler. (Browns)

A tractor front-end loader, or a telescopic fork lift equipped with one of a number of different types of big bale handling attachment, provides an inexpensive method of handling and stacking big round or big square bales. These attachments include a double bale spike for lifting and carrying either two round bales or one big square bale. A bale gripper, which hydraulically squeezes big round or big square bales, rather than penetrating them with a spike, can also be used for this work. Some have hydraulically operated side gripper arms for handling wrapped silage bales, while others have

Plate 16.20   A bale gripper for handling wrapped round silage bales. (Ritchie)

Plate 16.19   After reversing this bale transporter up to a previously built stack of bales, hydraulic rams are used to raise the body to a vertical position and grip the bales. The transporter is then lowered and the bales are taken to the stack.

Plate 16.21   The bale collector on this big square baler leaves the bales in groups of three for later collection.

hydraulically operated claws and can handle up to three big square bales at the same time.

Various types of self-loading trailer provide another method of automatically collecting and transporting bales. The hydraulically operated self-loading big square bale trailer (Plate 16.22) picks up single bales and stacks them on the trailer. The bales are unloaded by lifting the trailer floor to a vertical position and then driving away to leave the bales in a stack, either in the field or at the farm. Similar automatic self-loading trailers can be used to load and transport big round bales from field to farm. Stacking at the farm is usually done with a telescopic handler equipped with either a round or square bale grab or a spike.

Plate 16.22   This bale collector picks up and loads big square bales and then stacks the load in the field or farmyard. (Transtacker)

Plate 16.24   A control box in the tractor cab, with a separate switch for the hydraulic ram on each of the eight loading arms, is used to load or unload the bales on this round bale transporter. (Ritchie)

Plate 16.23   Spikes on these hydraulically operated bale lifting arms allow them to pick up big square bales one at a time and load them on to the transporter. The lifting arms are also used to unload the bales at the stack. (Ritchie)

Plate 16.25   A load of round bales being hauled to the farmyard. (Massey Ferguson)

CHAPTER 17

Combine Harvesters

Combine harvesters are used, with various types of header or cutter bar, to harvest a wide range of crops including cereals, grass and clover seed, peas, beans, oilseed rape, linseed and maize. A combine cuts the crop, conveys it to the threshing mechanism and then sorts the grain or seed from the straw and chaff. The straw is either left in a swath on the ground behind the combine or chopped into short lengths and spread on the stubble. The chaff is blown from the back of the combine and the grain is elevated to the grain tank (hopper). A high capacity auger unloads the grain tank into a trailer pulled alongside the combine in about two minutes.

Size and type of farm will determine choice of combine. Mixed farms will usually have a harvester with a cutting width of 4.5 m (15 ft) to 6.6 m (22 ft). Combines on large arable farms may have a cutting width of 7.5 m (25 ft) or 9 m (30 ft) capable of harvesting 800 ha (2,000 acres) or more in a season. Even wider cutter bars, with a 12 m (40 ft) cutting width are used on some very high output combine harvesters.

Most cutter bar tables have to be detached and mounted on a special trailer before they can be towed on the public highway by the combine or a tractor.

Depending on size, combine harvesters have an engine in the 150–450 kW (200–600 hp) bracket. With a few exceptions on older or smaller models, all combines have a hydrostatic transmission with a stepless range of forward and reverse speeds including a 'fast' road speed, and hydrostatic steering. Some have four-wheel drive, others have tracks instead

Plate 17.1   This 35 ft (10.5 m) cut combine harvester with a 496 hp engine has a cylinder and concave threshing unit and twin separating rotors. (Massey Ferguson)

Plate 17.2   Wide cutter bars have to be towed on a trailer behind the combine when travelling on the public highway. (John Deere)

of the main driving wheels. Almost all combines now have an air-conditioned cab with touch screen and switch- or dial-based controls, replacing the levers on older machines.

Plate 17.3   Rubber tracks with independent suspension and a large ground contact area reduce soil compaction and increase traction when combining across a slope in wet conditions. (John Deere)

Some combines have automatic steering provided by GPS or laser-guided navigation.

CROP FLOW

There are three stages in the flow of the crop through a combine harvester:

1. Cutting and elevating
2. Threshing
3. Separating and cleaning

Cutting and Elevating

A standard cutter bar with a knife, reel and an auger is used to harvest most combinable crops, while crops such as maize require a different header. A divider at each side of the cutter bar separates the standing grain from the crop that will be cut on the next pass. Grain lifters can be attached along the width of the cutter bar to lift laid or tangled crops and avoid losses in crops where the ears are close to the ground.

A reciprocating knife, with sections or blades supported by fingers bolted to the cutter bar frame, cuts the crop. A ledger plate on each finger creates the stationary cutting edges which provide a scissor-type cutting action. Knife clips hold the sections close to the ledger plates to give a clean cut. The knife sections have serrated cutting edges, which are ideal when harvesting cereal crops, and they do not have to be sharpened. When there is a lot of green material in the stubble, it may be necessary to use a smooth sectioned knife.

The crop is cut while it is held against the knife by the reel, which has spring steel tines attached to the reel bars. The tine angle can be

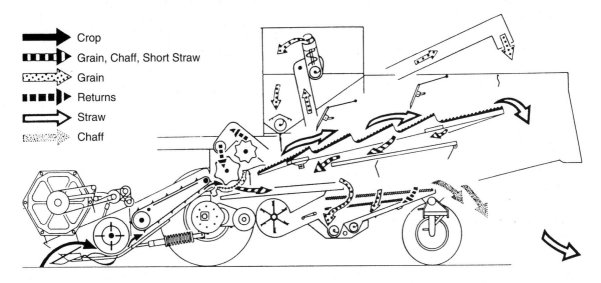

Key:
- Crop
- Grain, Chaff, Short Straw
- Grain
- Returns
- Straw
- Chaff

Figure 17.1   Crop flow through a combine harvester with straw walkers.

Plate 17.5   Vertical reciprocating knives on both sides of the cutter bar help to separate tangled stems when combining oilseed rape and some other tall crops. (Claas)

Plate 17.4   The knife cuts the crop, which is placed on the cutter bar table, where an auger with left- and right-handed flights carries it to the centre of the table and retracting tines feed the crop to the main elevator. (Massey Ferguson)

adjusted to improve cutting and intake efficiency in laid or lodged crops. On most combines, the height of the reel is adjusted with hydraulic rams controlled from the cab; a hand lever is used on some older machines. Reel speed, normally up to 50 rpm, is controlled from the cab by an infinitely variable hydrostatic drive. Older combines may have an adjustable chain drive to the reel.

A vertical reciprocating knife can be fitted to one or both sides of the cutter bar, especially when harvesting tall crops such as oilseed rape and field beans, where they act as dividers to separate tangled crops, and help to reduce seed losses and improve output.

After it has been cut, the reel moves the crop on to the cutter bar table or platform, where an auger with both right- and left-handed flights conveys it to the centre. Here, retracting tines on the auger feed the crop to the main elevator. The main elevator, normally of chain and slat construction, lifts the crop in an even flow to the threshing cylinder or threshing rotor. An uneven feed may either overload or starve the threshing unit, which in turn reduces the efficiency of the straw separating mechanism and grain sieves. Most combine harvesters have a reversing mechanism in the main elevator drive which can be used to clear blockages.

The cutter bar table is able to float, allowing the knife to follow the ground contours and achieve an even length of stubble. Many combines have an automatic cutting height control that maintains a consistent stubble height regardless of cutter bar width or uneven fields. A typical height control system uses hydraulic rams linked to mechanical sensors, which maintain a pre-set cutter bar ground pressure and hold the cutter bar at the required height.

Some combine harvesters have an automatic table levelling mechanism that keeps the knife level with the ground when working on uneven or sloping land (Plate 17.10). Sensors at each side of the cutter bar table, which follow the ground contours, actuate hydraulic rams used to tilt the table up or down by a few degrees on either side of the combine. One example of an automatic table levelling system tilts the cutter bar so that either side can be up to 320 mm higher than the opposite side.

A stone trap collects solid objects such as large stones which find their way on to the cutter bar platform, thus helping to prevent damage to the threshing cylinder. The stone trap, which may be situated under the front beater or in front of the main elevator, must be emptied regularly. A trapdoor is provided for this purpose. Combines with a longitudinal threshing rotor have a stone retarding roller or deflector plate to prevent stones entering the threshing mechanism.

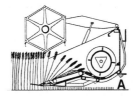

Figure 17.3   Types of cutter bar table. (Massey Ferguson)

The distance between the cutter bar and table auger can, especially in lodged crops, affect the way the material is collected by the auger and fed to the elevator. Although this distance (Figure 17.3A) varies on different models of combine, on some cutter bars it can be varied according to the crop to achieve a more efficient head-first feed to the threshing cylinder or rotor. Another design of cutter bar table (Figure 17.3B) has a power-driven endless conveyor between the knife and the auger, which gives more space for a head-first feed in all conditions and helps to even out the crop flow to the elevator and threshing cylinder.

Threshing

Combine harvesters have either a cylinder or concave or a longitudinal rotary threshing unit where the grain or seed is rubbed from the ears or pods.

Cylinder and concave threshing The main elevator carries the crop up to the threshing cylinder (drum) and concave, which consists of a heavy rotating cylinder with rasp-like beater bars. The concave is a curved frame with a set of stationary rasp bars under and very close to the cylinder. A perforated grating between each of the concave rasp bars allows the threshed grain to fall through it onto a grain pan or tray under the concave. The grain is rubbed from the ears by the rasp bars as the crop passes

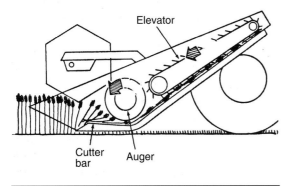

Figure 17.2   The cutter bar table and elevator. (Massey Ferguson)

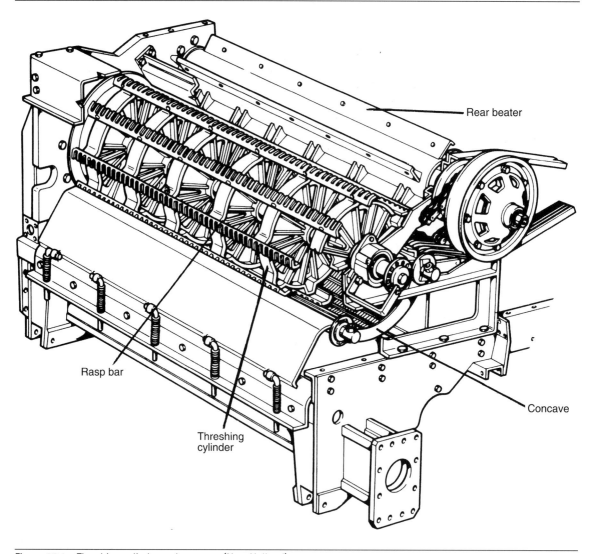

Figure 17.4   Threshing cylinder and concave. (New Holland)

between the cylinder and concave. In good threshing conditions, at least three-quarters of the grain will fall through the concave onto the grain pan along with some chaff and short lengths of straw, which will be separated from the grain on the sieves.

Depending on the size of the combine, the threshing cylinder is usually between 450 and 650 mm (18 and 26 in.) in diameter, and between 1.1 and 1.7 m (43–67 in.) wide, with eight or ten rasp bars. As the cylinder diameter varies with different models, the threshing speed of the cylinder is given in m/min. This is the peripheral speed of the cylinder and refers to the distance travelled, in this case by the rasp bars, in one minute. Peripheral speed in metres per minute can be calculated by multiplying the cylinder diameter (m) by its rpm (revolutions per minute) and multiplying the result by 3.142.

Plate 17.6   A sectional view of a combine harvester with a cylinder and concave threshing unit and straw walkers. (Claas)

When harvesting cereals, the cylinder speed will be about 1,800 m/min. This is equivalent to 6,000 ft/min, which means that the crop passes between the cylinder and concave at a speed of almost 70 mph. The cylinder speed for threshing dried peas is about half of that used to thresh cereal crops.

Combine harvester instruction manuals give the threshing speed and concave setting for different crops and harvesting conditions. Both of these adjustments are made remotely from the cab on many combines, but spanners may be needed to change the cylinder speed and concave clearance on some older models.

Combines with a cylinder and concave may have either a front or a rear rotary beater; some have both. A front beater speeds up the crop on its way to the threshing cylinder, and on some combines it is also a pre-threshing cylinder and concave which partly threshes out the grain before the crop reaches the main threshing cylinder. The rotary rear beater slows the crop after it has been threshed to ensure maximum separation of any remaining grain.

Different cylinder and concave threshing system layouts used on combines with straw walkers are illustrated in Figure 17.5. The pre-threshing cylinder and adjustable concave (A) speeds up the crop and provides an even feed to the main threshing cylinder. It also threshes out and separates some of the grain from the straw before it reaches

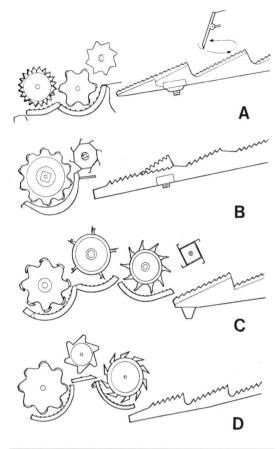

A

B

C

D

Figure 17.5   Cylinder and concave threshing layouts with various beater arrangements.

the main cylinder. After the crop has been threshed, the straw is moved on to the straw walkers by the rear beater, and the grain is collected on the grain pan. The pre-threshing cylinder runs at about 80 per cent of the main cylinder speed.

The dual drum arrangement (B) consists of a large primary drum and concave, with a second threshing cylinder that feeds the straw and any remaining grain on to the straw walkers. Any grain shaken from the straw walkers joins grain from the concave on the grain pan.

When it has been threshed by the main cylinder and concave (C), the rear beater directs the crop down to the rotary separator, which accelerates the straw to remove any remaining grain before it passes on to the straw walkers.

After passing through the threshing cylinder (D), a beater feeds the crop to a rotary separator, where more grain is removed before it moves onto the straw walkers.

All combines have a returns system to deal with partly threshed ears, etc, which are too large to pass through the grain sieves. They are either returned to the main threshing cylinder for re-threshing or elevated to a re-thresher

Plate 17.7   Detail of the inclined threshing and separation rotor in a single rotor combine. (Case IH)

unit for further treatment and then returned to the grain sieves.

Rotary threshing The principle of rotary threshing was invented almost 200 years ago. Rotary or axial flow combines differ from cylinder and concave machines in that they use one or two axial rotors to thresh and/or separate the grain from the straw. A rotary combine is shorter than a straw walker machine with a comparable output because it has a combined threshing and separating mechanism. The claimed advantages are a higher working speed and gentler handling of the crop. Some designs, often referred to as 'hybrid' machines, have a cylinder and concave to thresh the crop with one or two rotors to separate any remaining grain from the straw.

Another design uses a single axial rotor for both operations. Here, the inclined threshing rotor runs the full length of the combine body. Depending on model, the threshing rotor will be between 1.5 and 3.0 m long and 600 to 800 mm in diameter. An impeller at the intake end of an enclosed rotor gathers the crop from the elevator and accelerates it rearwards in a spiral movement around the threshing rotor. Rotor speed and concave clearance can be adjusted for different crops and harvesting conditions.

The rotor threshes the crop against a fixed concave grate around the lower part of the rotor as it moves towards the back of the machine. The threshed grain is thrown outwards through the concave to a conventional grain sieve and fan cleaning system. A discharge beater at the

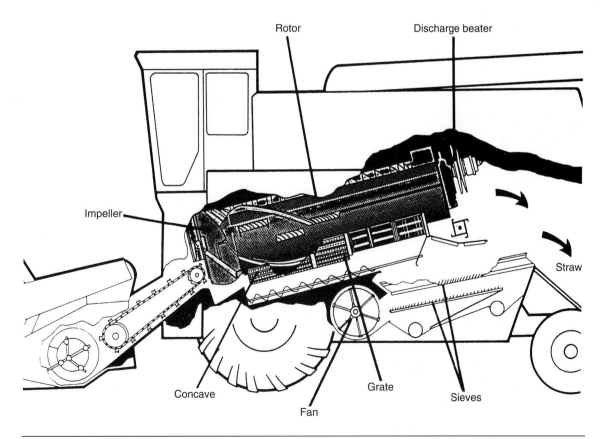

Figure 17.6   Axial Flow combine. (Case IH)

outer end of the rotor extracts any remaining grain from the straw before it leaves the back of the combine.

An alternative rotary threshing system has two longitudinal threshing rotors and concaves. With a twin-rotor system, the crop is divided into two separate streams as it leaves the main elevator and is fed into the contra-rotating threshing rotors. The first section of each rotor threshes the crop and the grain falls through the concaves on to the sieves. Any grain still in the straw after threshing is removed as it passes along the second part of the rotors and falls through the concaves on to the sieves. A beater at the end of the rotors discharges the straw from the back of the combine. Rotor speed and concave settings are controlled from the cab, and a sensor on the beater measures grain losses from the back of the machine.

Plate 17.8   Sectional view of a combine harvester with a rotary threshing system.

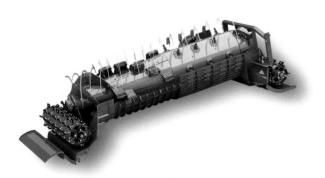

Plate 17.9   Detail of the threshing rotor used in the combine harvester shown in Plate 17.8. (John Deere)

Separation and Cleaning

Combine harvesters with a cylinder and concave threshing system separate the grain from the straw through the action of the rotating cylinder against the fixed concave. Most of the threshed grain falls through the concave and is gathered on a grain pan under the concave. The straw passes on to the straw walkers, where their shaking action removes any remaining grains.

Machines with a rotary threshing system drop most of the threshed grain through a concave on to a grain pan. Any remaining grains are removed by the rotary separating system as the straw passes through the combine between it and the concave, which runs the full length of the machine.

All combine harvesters have grain sieves and a fan under the threshing unit which separate partly threshed heads, small grains, weed seeds and chaff from the good grain which is elevated to the grain tank. The partly threshed heads are removed from the sieves and returned for further treatment. The chaff and small seeds on the sieves are blown from the back of the combine.

Straw Walkers

After threshing, the straw passes on to the rows of straw walkers with a series of steps which shake the straw. This action removes any loose grains as it travels towards the back of the combine. The grains are collected in troughs that form the bottom of the straw walkers, or by a full-width grain tray under them, and these pass it forwards to the grain pan under the concave. Depending on size, a combine may have four, five or six rows of straw walkers mounted on two crankshafts. Some have one or more canvas curtains hanging down over the straw walkers to slow down the straw as it moves towards the back of the machine. Others have mechanical agitator tines above the straw walkers, which shake the straw and retard its progress towards the back of the machine.

Figure 17.7   Examples of agitator tines, which stir the straw as it moves along the straw walkers.

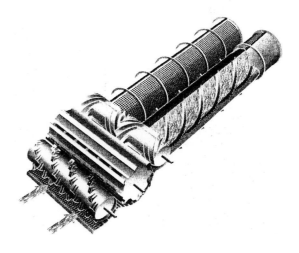

Figure 17.8   Rotary separation system with twin rotors. (Claas)

Rotary Separation

Some combines, including the axial flow type of machine, have a rotary separation system which removes any grain left in the straw after the crop has been threshed. The rotary separation system shown in Figure 17.8 has two rotors with auger profiles which remove any loose grains from the straw as it moves towards the back of the combine. An impeller divides the straw into two equal streams as it leaves the threshing cylinder and directs them to the two rotors housed above full-length concaves. Depending on the crop and harvesting conditions, the rotors run at a speed of 500–1,000 rpm and remove any remaining grain which falls through the concaves on to a grain pan and then on to the sieves.

A similar twin-rotor separation system on another type of rotary combine has two counter-revolving tine rotors which receive the crop from an overshot rotary beater drum behind the main threshing cylinder. The 3.4 m long separator unit can, depending on the crop and conditions, be set to run at a speed of between 570 and 740 rpm. The separating tines remove any remaining grain from the straw, which is passed through the concaves as it moves towards the back of the machine. Filler plates can be fitted to the rear sections of the concaves to prevent large amounts of chaff falling through with the grain on to the sieves.

Another rotary separation system has a full-width full twin flow rotor and two rotary discharge beaters instead of straw walkers. Most of the grain is removed from the straw by the main threshing cylinder, rear beater and rotary separator (Figure 17.5C). The straw and remaining grain then passes to the twin flow rotor and concave, which splits the crop sideways. The rotor spins the straw for one revolution while centrifugal force presses it against the concave to extract any loose grains. Grain from the centre section falls on to the sieves, and augers lift the grain from the side sections on to the sieves. The rear discharge beaters pass the two streams of straw along the sides of the combine; they are then combined and either left in a swath or directed to the straw chopper.

Cleaning the Grain

The grain sieves sort the grain and chaff from the concave and straw walkers or other separation system before the good grain is elevated into the tank. The grain separation unit (Figure 17.9) consists of two sieves, one above the other, and a fan. The shaking action of the sieves separates the good grain from the partly threshed heads, short pieces of straw, weed seeds and chaff. A fan provides a constant flow of air to keep the sieves free from blockages.

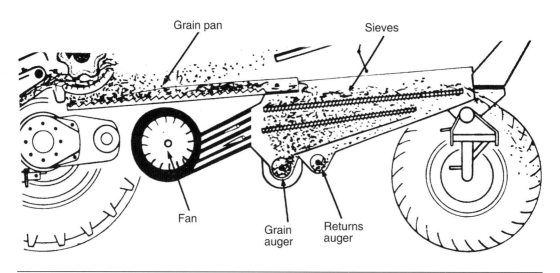

Figure 17.9   The layout of the grain sieves, fan and returns elevator. (New Holland)

The top sieve (chaffer sieve) allows every-thing through except pieces of straw and partly threshed ears. Most combines have an exten-sion at the back of the top sieve with larger holes. These allow any partly threshed material to fall through to the returns auger.

The bottom sieve (grain sieve), which receives the material from the top sieve, sep-arates the good grain from the weed seeds, small grains and chaff. Some combines have a grain pan under the top sieve which directs the material to the front of the bottom sieve. The good grain, together with any remaining small grains, passes through the bottom sieve and is conveyed by an auger to the grain elevator, which lifts it to the grain tank.

A second auger – the returns auger – collects any partly threshed ears which pass over the back of the top sieve and carries them to the returns elevator. On some combines, the returns are lifted to the main cylinder for re-threshing; others put them back on the top sieve for further sorting. Depending on the condition of the returns, some combines offer the choice of either directing the returns to the cylinder or top sieve. Another design has a re-thresher unit built into the returns elevator, and after re-threshing the returns are spread on the grain pan or sieves.

The fan, an important part of the separa-tion system, keeps the sieves clear by blow-ing chaff, weed seeds, etc from the back of the combine. Depending on harvesting con-ditions, the fan can be set to blow air through the front, middle or back of the top and bot-tom sieves.

Combining across sloping land will affect straw walker performance. The grain will slide across the separating system and over-load one side of the sieves, which can result in excessive grain loss from the back of the machine.

Some combines have an automatic self-levelling mechanism for the straw walkers and sieves. Controlled with hydraulic rams, this

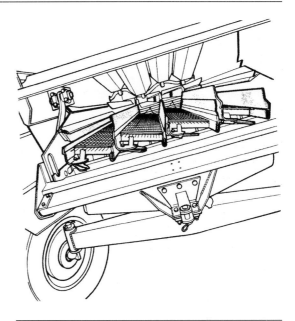

Figure 17.10   Self-levelling grain sieve. (New Holland)

arrangement keeps the sieves and straw walk-ers evenly loaded across their full width when working across sloping ground. Other com-bines have a hydraulic ram on both front wheel housings, used to raise or lower the wheels up and down in relation to the chassis to keep the machine level when working on slopes. This system ensures that the straw walkers and sieves remain level and are evenly loaded across their full width.

Another automatic control system keeps the cutter bar parallel with the ground when work-ing on undulating or sloping ground and is particularly useful on combine harvesters with very wide cutter bars.

The harvested grain is elevated to a tank or hopper situated at the top of the machine. It is usual to have an auger at the top of the elevator to ensure even filling across the full width of the tank. Grain tank capacities vary according to the size of the harvester. Typical grain tank capacities range from 8,000 litres on a 6 m cut machine up to 12,000 litres on a high output 12 m cut combine harvester.

Plate 17.10   The cutter bar table on this combine follows the ground contours when working across slopes, while the automatic self-levelling system maintains the combine body and cab in an upright position. (John Deere)

A high capacity unloading auger, which is swung into position over a trailer by a hydraulic ram, can empty an 8,000 litre grain tank holding about 6.5 tonnes or 250 bushels of wheat in about two minutes. Unloading can be done on most combines while cutting continues. A pressure-activated warning light, which flashes when the tank is almost full, alerts the driver who has the task of hauling the grain to the store that he must get to the combine as soon as possible.

Plate 17.11   A straw chopper is a standard attachment on many combine harvesters. When the straw is to be baled rather than chopped, the drive to the chopper rotors is disengaged and, after lowering the chopper unit, the straw is left in a swath behind the machine ready for the baler. (Claas)

ATTACHMENTS

When the straw comes off the straw walkers or rotary separation system, it may be left in a swath for baling or chopped into short lengths and spread on the stubble.

Straw chopper Many combines have a straw chopper attached to the back of the machine. It cuts the straw into very short lengths and spreads it over a similar width to the cutter bar. The chopper unit has a number of rows of high-speed swinging blades that cut the straw against a row of fixed knives. The cutting edges may be serrated and are usually double edged so that they can be reversed

Plate 17.12   Finely chopped straw and the chaff from the straw chopper are spread over a wide area behind the combine. (Claas)

when worn. The chopper is usually vee-belt-driven through a gearbox and controlled by a clutch. Deflectors on the chopper unit can be adjusted to change the direction and width of spread or to direct the straw away from the chopper and leave it in a swath ready for the baler.

Chaff spreader Normally driven by a hydraulic motor, the chaff spreader distributes the large quantity of chaff coming off the sieves which otherwise would form a thick mat on the stubble behind the machine. On combines with a straw chopper, the chaff is usually spread with the chopped straw.

CONTROLS AND ADJUSTMENTS

The combine harvester must be set to suit the crop being harvested. Setting the machine for threshing different crops and varying field conditions is controlled from the cab on many combine harvesters. However, on older machines, manual controls and the use of a spanner may be necessary when setting the machine or making adjustments during the working day.

There are considerable variations in the settings for cereals, beans, grass seed and other crops. The settings given below are only a guide; the instruction book, or computer monitor, must be consulted when preparing a combine for work. Once in the field, fine tuning can be made where necessary.

The Cutter Bar Table

Efficient threshing requires an even feed to the cylinder and concave or threshing rotor. Careful adjustment of the various components on the cutter bar table will help achieve this.

The Dividers

These should be set to part the crop about to be cut from that which will be cut on the next pass. Straw must not be allowed to build up around the divider points.

The Pick-Up Reel

The reel is either vee-belt-driven with variable speed pulleys or hydraulically-powered via an infinitely variable speed hydraulic motor.

When working in a standing crop, the following settings are a useful guide:

- The reel slats should touch the straw just below the ears.
- The reel speed should be slightly faster than the forward speed of the combine. As a guide, the reel should turn slightly faster than the combine front wheels. The speed is altered hydraulically from the cab with a switch or lever. Some combines have an automatic control which synchronises the reel speed with the forward speed.
- The reel is moved backwards and forwards with a hydraulic ram and the tine bars or slats should be set in front of the knife. The tines on the pick-up reel should be almost vertical when they come into contact with a standing crop.

When harvesting laid or tangled crops:

- Lower the reel.
- Increase the reel speed.
- Angle the pick-up tines towards the knife. In this position they will lift the crop to help give a cleaner cut and reduce losses.

The Knife

The cutter bar fingers and the knife must be in good condition:

- The knife sections must be undamaged and securely riveted to the knife back.
- When a smooth sectioned knife has to be used, it must be kept sharp.
- Adjust the knife clips to ensure that the knife sections are firmly held against the ledger plates, which provide the cutting edges on the fingers.
- Check that the knife is correctly registered. It should travel from the centre of one finger to the centre of the next, or the next but one, depending on the model of combine and the type of knife drive mechanism.

Height of Cut

This is altered with a lever or switch in the cab which controls the hydraulic rams used to lift and lower the cutter bar table. The usual cutting height is about 100 mm from the ground, but this is usually higher when combining oilseed rape. Many combines have an automatic table height control which maintains a pre-set stubble length.

In heavy crops, there is a risk of feeding too much material into the combine, especially with smaller output models. This will overload the threshing mechanism and straw walkers and grain will be lost from the back of the combine in the straw.

When it is desirable to reduce the volume of crop being taken into the combine, this should be done by reducing forward speed or raising the cutter bar a little. Never reduce crop intake by driving in a way that uses less than the full width of the cutter bar. This will result in an uneven feed to the cylinder and reduce threshing efficiency.

The Table Auger

This must be set to avoid blockages or straw wrapping around the auger flights. The auger flights must be at a suitable height for the crop being harvested. If they are too low, the flights will rub grain from the ears or pods. The clearance between the tips of the retracting tines and the platform can be adjusted to suit different crops and harvesting conditions. The tines must not touch the bottom of the cutter bar table.

Many combines have a reversing mechanism in the drives to the pick-up reel and table auger which are used to clear blockages on the cutter bar table and in the crop elevator.

The Stone Trap

This is usually sited at the top of the main elevator and should be emptied daily to ensure stones and other solid objects do not reach the cylinder and concave or threshing rotor.

The Threshing Unit

Careful adjustment of the cylinder and concave or threshing rotor and concave is important to ensure efficient threshing with minimum loss of seed or grain. Settings will depend on the type of crop, its condition and its moisture content. On many combines, electronic controls in the cab are used to make these settings, often via a monitoring screen which shows all of the machine's functions and checks threshing efficiency.

The threshing cylinder The speed of the threshing cylinder, or drum, should be as low as possible to give complete threshing. It is adjusted with a pair of hydraulically operated variable speed belt pulleys. Most combines have a stepless electro-hydraulic adjustment activated by a switch or lever in the cab. Cylinder speed is registered on a dial or screen.

A typical cylinder speed for threshing cereal crops is 1,800 metres per minute. This is the speed at which the crop passes between the cylinder and concave. For example, a 60 cm diameter threshing cylinder achieves this speed when running at 950 rpm.

Low cylinder speeds will not thresh all of the grain from the ears. High speeds will crack the grain and, in very dry conditions, break up the straw, which may find its way into the grain tank and spoil the sample.

The concave The clearance between the cylinder and concave should be set as wide as is required to give complete threshing. A wide clearance will leave some unthreshed grain in the ears. The grain will be damaged if the concave is set too close to the cylinder. The clearance is always greater at the front than at the back and this difference is automatically retained at all concave settings.

The concave clearance is adjusted with a switch from the cab on many combines, and the clearance is shown on a display panel in the cab. The concave clearance on some combines is adjusted manually. A typical concave clearance for threshing wheat is 10 mm at the front and 4 mm at the back. The concave is set very close to the cylinder when threshing small seeds such as grass and clover.

Filler plates can be placed between the concave bars when threshing small seeds to improve threshing performance. This prevents seed falling through the concave on to the grain pan and separation of the seed takes place on the straw walkers.

Rotary threshing A variable speed vee-belt drive combined with a high-low range gearbox is used to adjust the rotor speed to suit the crop and harvesting conditions. Depending on the model and the crop being harvested, rotor speed is from about 250 rpm for peas up to 1,250 rpm when harvesting cereal crops. The concave clearance is adjusted from the cab with an electric motor. Air drawn into the threshing unit by the intake impeller blows most of the dust from the back of the machine.

The straw walkers No adjustments are required, but it is important to keep them clean. Blocked straw walkers will prevent grain reaching the sieves and this grain will be lost in the straw as it falls from the back of the combine. Drive belts must be kept at the correct tension. A slack drive belt will slip, reduce the shaking effect and result in grain loss.

The sieves Most combine harvesters have adjustable top and bottom sieves. The size of the sieve openings is altered either with hand levers or electrical actuators controlled from the cab. Some combines have a set of round hole sieves, each with different diameter holes to suit a range of crops.

Adjustable sieves should be set as wide open as possible to obtain a good sample of grain. Losses will occur if the sieves are closed too much, and there will be pieces of straw and chaff mixed with the grain if the sieves are too open.

A typical top or chaffer sieve setting will have an opening of about 10 mm for cereals.

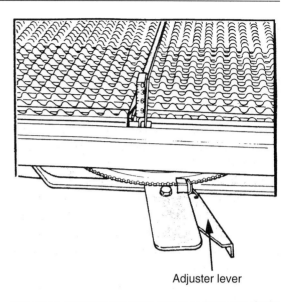

Figure 17.11   Detail of an adjustable grain sieve. (New Holland)

Settings for other crops will depend on the size of the seed.

The bottom or grain sieve will be set at approximately 7 mm for cereals. Some farmers use a round hole bottom sieve. A 16 mm diameter round hole sieve is suitable for cereal crops, and grass requires a 2.5–4.5 mm diameter sieve. A 5 mm diameter sieve is recommended for oilseed rape. This will give a better sample by preventing small pieces of pod and stalk which fall through the top sieve making their way into the tank.

The disadvantage of round hole sieves is the need to have a selection of sieves with different diameter holes for a variety of crops. However, round hole sieves tend to give a cleaner sample, as short pieces of straw do not readily pass through the holes.

The fan Adjustments are provided to control the direction and quantity of air. Deflector plates in the fan outlet can be set to direct the air to the front, middle or back of both sieves. The volume of air is controlled by the speed of the fan. Some combines have an adjustable air inlet into the fan, which provides an additional control of the airflow. Many combines have a variable speed drive to the fan, which is controlled from the cab. A typical variable speed fan speed has a range of 380 to 1,200 rpm.

Light seeds need a low volume of air, which is directed at the front of the sieves. Cereals need a greater flow of air directed to the centre of the sieves, while peas and beans require large quantities of air to keep the sieves clean. As a general guide, use as much air as possible without blowing seed from the back of the combine. Too little air will result in blocked sieves.

Combine Controls

There are many switches, levers and instruments in a modern combine harvester cab, both for driving the machine and for adjusting and controlling the cutting, threshing and separation mechanisms.

The cab has a full set of dials and instruments or a touchscreen on many modern machines which register threshing and separating efficiency, cylinder speed, concave clearance, fan speed, cutting height, etc. Other instruments monitor engine speed and performance, oil pressure, cooling system temperature, forward speed and hours worked.

Many combines have electronic monitoring and control systems in their air-conditioned cabs. A computer terminal stores the combine instruction manual and monitors grain loss, records crop yields, the area harvested, output per hour, etc. This information is displayed on

Plate 17.13   The cutting, threshing and driving controls and performance monitoring screen are all close at hand in the cab. (John Deere)

a screen and some of it can be printed for a permanent record. Some computer systems can be programmed to reset the machine for harvesting different types of crop. More sophisticated systems automatically adjust the combine settings as crop conditions change throughout the day, measure crop moisture content of the harvested grain or seed and produce a yield map for each field. Reel, cylinder and fan speeds, the grain tank unloading auger, cutting height, reel position, etc, are all controlled from the cab, either through the terminal with electronic switches or, on older machines, levers and pedals, which operate small electric motors or hydraulic rams.

The engine must run at maximum speed when combining to provide sufficient power to ensure efficient harvesting performance. A single joystick with a number of switches usually selects the speed range and direction of the hydrostatic transmission, providing a stepless speed. Some combines have an autopilot system, guided either by a GPS receiver or by a laser, which detects the top edge of the standing crop and automatically steers the combine to use the full width of the cutter bar.

A switch or pedal in the cab allows the driver to stop the drive to the cutter bar, reel, auger and elevator instantly if any of them are blocked or an object ahead of the combine is likely to find its way on to the cutter bar table. Many combines have a reverse drive for clearing cutter bar, auger or elevator blockages.

YIELD MAPPING

Computer technology has made it possible to measure crop yield on the move, and when linked to a GPS system, this can be used to make accurate yield maps. Crop yield is measured by monitoring the flow of grain into the tank, and signals from the grain flow monitor are sent to a computer in the combine cab and displayed on the screen in tonnes per hour or tonnes per hectare.

Plate 17.14   A sensor on each side of the cutter bar, linked to an automatic steering system, detects the edge of a standing crop to ensure its full working width is used. GPS signals provide an alternative system of steering guidance, and can also be used, in combination with a grain yield monitor, to make a yield map for the field. (Claas)

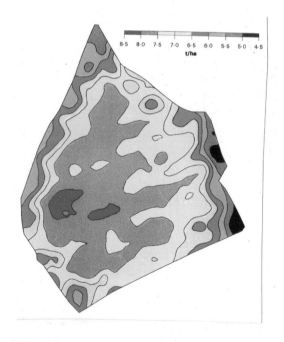

Plate 17.15   A yield map is made through a combination of the data obtained from the combine elevator's grain flow monitor and the machine's GPS system, which continuously maps its position in the field. (Massey Ferguson)

Global Positioning System (GPS) satellites are used to continuously record the position of the combine in the field and this, together with the yield information, is recorded by the in-cab computer. This information is transferred to the farm office and is used to make a map showing yield variations across the field. After two or three years, distinct yield patterns emerge and decisions can be made concerning seed rates and crop treatments. The yield map can be transferred to a similar computer terminal in the tractor cab, and by linking up with the GPS system, seed, spray chemical and fertiliser application rates can be automatically varied to match the yield potential on different parts of the field.

MAINTENANCE

- Regular cleaning during the season in important. The concave, straw walkers and sieves must be kept clean – if they are blocked with chaff, grain will be lost from the back of the machine. Combining in damp conditions often leads to clogged cylinder rasp bars and a blocked concave.
- The stone trap, when fitted, should be emptied daily.
- Correct tension must be maintained on all belt and chain drives.
- The elevators, which convey the crop to the threshing cylinder, the grain to the grain tank and the returns for re-threshing, should be checked regularly for damage and correct chain tension.
- The cutter bar should be checked for broken knife sections and damaged fingers. The knife must be a close fit on the fingers, and when a plain-sectioned knife is used it must be sharp.
- Thorough and regular lubrication of grease nipples is vital, but many combines have either sealed bearings that require no attention or an automatic lubrication system linked to most of the bearings on the combine.

- Check oil levels in the engine, transmission system, hydraulics and other gearboxes at regular intervals. Top up with the correct grade of lubricant when necessary.
- The engine has to work in very dusty conditions, which makes daily servicing of the air cleaner an essential part of the maintenance schedule. A blocked air cleaner will restrict airflow into the engine and reduce power output. The cab air filter also requires daily cleaning.

STORAGE

The combine harvester is a very expensive machine with a short working season. Careful storage between harvests will help to ensure a long and efficient life and keep corrosion at bay. Points to remember include:

- Clean the combine thoroughly, removing all traces of grain, straw and rubbish, which may attract rats and mice.
- Lubricate thoroughly and check oil levels.
- Protect bright parts with rust preventative.
- Service the engine, drain the cooling system if there is no anti-freeze in the water and keep the battery well charged as a discharged battery will freeze. It is a good idea to seal the air cleaner intake and engine breather to prevent condensation inside the engine.
- Check tyre pressures and inflate as necessary.
- Check the combine for damage, and replace any broken or worn parts.
- Store under cover, leaving elevator doors open to discourage rats and other vermin from taking up winter residence in the combine.

USING COMBINE HARVESTERS

The combine harvester presents few driving problems in the field; the skill is setting the cutting and threshing mechanisms, and even this is a simple matter on combines with electronic controls which will set up the machine for different crops and working conditions.

The headland is cut first: three to five rounds, depending on the width of cut, will provide ample turning space. If the crop on the headlands is a little damp when opening up a field, the extreme edges of the field can be left until later in the day when they have had a chance to dry. Care should be taken on the first headland round if there are overhanging trees that might damage the grain unloading auger. This can be avoided by keeping the unloading auger spout folded while cutting the first round.

With the headland cleared, the field may either be cut round and round or in lands (sections). The method will depend on the size and shape of the field and the availability of trailers for carting the grain to the store.

Drive the combine to ensure an even feed to the threshing cylinder or rotor by using the full width of the cutter bar. In heavy crops, reduce forward speed or increase the height of the stubble if the combine is being overloaded. Never reduce the cutting width because, apart from being wasteful, the cylinder, sieves and straw walkers may not be evenly loaded over their full width and this may result in damage to the grain.

Safety Devices

Many combines have a performance monitor in the cab, which as well as monitoring the operation of the various harvesting systems also warns the driver of faults in the engine, main drives, cutter bar and threshing mechanisms.

Slip clutches protect the main drive systems – including reel, table auger, main elevator and also the grain and returns elevators – from damage through overloading or blockages.

Vee-belt drives to the straw walkers and sieves should be tensioned to allow belt slip if severe overloading occurs.

THRESHING PROBLEMS

Grain loss from the straw walkers or sieves and damaged grain are the main problem areas.

It is a simple matter to locate grain losses from combines with an in-cab monitoring system. The monitor screen is connected to sensors which measure the amount of grain falling through the straw and chaff as it leaves the straw walkers and sieves. Other sensors measure the operating speed of the threshing cylinder or rotor and the fan. The driver only needs to press a button, or touch the monitor screen, to obtain this information and, after making minor adjustments, the results can be checked.

When using a combine without a performance monitor, the driver can check for grain loss by looking for grains in the straw behind the combine. Grain losses can occur in front, underneath or behind the combine. To check if grain is being lost, drive the combine for a short distance, stop the machine and carry out the following checks:

1. Look in front of the combine. Loose grain on the ground in front of the cutter bar in a standing crop suggests it is shedding grain from the ears because it is over-ripe.
2. If there is no shedding, look on the cutter bar platform. Loose grains here may be because:
   a. The reel speed is too fast.
   b. The reel is set too high so that it knocks grain from the ears.
   c. The auger flights are too close to the platform floor. In very ripe crops, this can result in grain being rubbed from the ears.
3. Assuming there are no grain losses at the front of the combine, look under the machine for signs of leakage from, for example, an ill-fitting elevator door.
4. The final checks are made behind the combine. Shake some straw from the swath over a sheet to see if any broken ears or loose grains fall from the straw. The presence of either indicates:
   a. A wide concave or clearance or low cylinder speed leaving unthreshed heads in the straw.
   b. Blocked concave bars resulting in loose grains left in the straw.

c. Blocked straw walkers or a slack drive belt.

d. Overloaded straw walkers because too much straw is entering the machine. Either slightly reduce the forward speed or raise the cutter bar slightly.

If there is no grain in the straw but loose grains are found in the stubble underneath the swath, the loss will be from the sieves. Likely causes include:

a. Sieves closed too much.

b. Blocked sieves.

c. Insufficient air from the fan, or the deflectors are angled too far towards the back of the sieves.

5. Damaged grain in the tank must also be treated as a loss because it spoils the quality of the sample. Cracked grains in the tank may be due to:

a. Threshing cylinder or rotor speed too high.

b. The concave being set too close to the cylinder or rotor.

c. Failure to keep the combine fully loaded. The straw cushions the grain as it is threshed. If there is too little straw in the combine, grain may be damaged. The problem can occur when starting into work or at the end of a bout.

d. Excessive returns – grain and partly threshed ears which pass over the top for re-threshing – resulting in damaged grain. If the top sieve is not blocked, then it should be opened slightly to allow more material to fall through on to the bottom sieve.

GRAIN STRIPPERS

A stripper header is an alternative to the cutter bar, auger and reel. It removes the ears from standing crops and feeds them into the combine to be threshed in the normal way. The straw is left standing to be dealt with later, when it may either be chopped or cut and baled. A major advantage is a much higher throughput, and crops can also be harvested when the grain is ripe but the straw is not ready to be harvested.

Plate 17.16　Combining barley with a stripper header in place of the conventional knife and reel header. (Shelbourne Reynolds)

Plate 17.17　The stripper header rotor is fitted with rows of keyhole-shaped teeth through which the crop heads pass to remove the grain. (Shelbourne Reynolds)

The stripper header has a rearwards rotating octagonal rotor with eight rows of keyhole-shaped plastic teeth that strip the ears and small pieces of leaf from the standing crop. The grain, chaff and minimal amounts of leaf are conveyed by an auger and elevator to the threshing cylinder. The stripper rotor speed varies from 400–1,000 rpm, depending on type of crop and harvesting conditions. Stripper headers, which range in widths from 3–13.5 m, are attached to the combine with the same quick-attach mechanism used for a cutter bar table. They tend to be more suited to countries where cereals have a low volume of straw.

CHAPTER 18

Root Crop Machinery

Plate 18.1 Features of this self-propelled sugar beet harvester include a 444 kW (604 hp) engine, hydrostatic drive to all six wheels and a discharge conveyor that empties the 27 tonne capacity tank in one minute. (ROPA)

Plate 18.2 A 110 kW (150 hp) tractor is required for this six-row trailed sugar beet harvester with the topper mounted on the tractor's front linkage/PTO. (Grimme)

SUGAR BEET HARVESTERS

Most the sugar beet crop grown in Great Britain is harvested with self-propelled machines that lift six, nine or twelve rows in a single pass across the field. Self-propelled harvesters are either owned by farmers, farming groups or agricultural contractors who harvest the sugar beet crop grown on farms in their locality. Some farmers with a small acreage of sugar beet use tractor drawn harvesters.

Trailed Sugar Beet Harvesters

Tractor drawn side-delivery and tanker sugar beet harvesters require a tractor in the 60–112 kW (80–150 hp) bracket. Depending on model, they top, lift and clean between two and six rows in a single pass, and either side-elevate the crop into a trailer or hold it in a tank, or bunker, on the machine. When the tank is full, the crop is either elevated into a trailer pulled alongside the harvester, or the harvester is taken to the headland, where the tank is emptied onto a heap for later delivery to the factory.

There are two types of trailed harvester. Single-stage harvesters top, lift, clean and elevate the crop into a holding tank on the machine or side-elevate it into a trailer. Two-stage harvesters consist of a tractor drawn topper followed by a second machine that lifts and cleans the crop and side-elevates it into a trailer. Some machines also have a small holding tank.

Plate 18.3   This six-row topper has a flail rotor and scalper blades. (Grimme)

Self-Propelled Harvesters

Depending on size, a self-propelled sugar beet harvester has an engine in the 225–490 kW (300–650 hp) power range, hydrostatic transmission, four- or six-wheel drive and two- or three-axle steering. Most six-row machines have a tank holding between 12 and 30 tonnes, and it takes about one minute to empty the tank. The largest nine- and twelve-row models, with an engine of up to 450 kW (600 hp), six-wheel drive and six-wheel steering, have very wide tyres to reduce the damage done to the soil structure.

Plate 18.4   Sugar beet, topped with the six-row mounted topper (Plate 18.3), ready to be harvested. (Grimme)

A twelve-row model can harvest 100 tonnes of sugar beet or more in an hour.

After a front-mounted flail defoliator has removed most of the leaves and moved them sideways onto cleared land, the roots are topped, lifted, cleaned and then elevated into the holding tank. A high-capacity side-delivery elevator is used to discharge the harvested crop, either into a trailer or onto a clamp at the headland.

Most of the controls and adjustments on a self-propelled harvester, including lifting depth, topping height, speed of the cleaning mechanism and engaging the drive to the tank discharge elevator, are made from the cab. After making the first run to set up the machine, the in-cab electronic guidance system keeps the toppers and lifters on the rows and gives the driver time to check harvester performance. The cab on most multi-row harvesters has a full range of fingertip controls and electronic monitoring systems, and gives the driver a clear view of the toppers, lifters and cleaning system. A multi-function joystick is used to select cruise control and manual or automatic steering to keep the harvester on the rows. The joystick is also used to set topping height, lifting depth and regulate the cleaning mechanism. Once set, the in-cab computer, linked to sensors on the topping, lifting, cleaning and unloading systems, monitors and automatically adjusts them to give maximum output in changing harvesting conditions. The computer also records the area harvested and operational data.

Toppers

Self-propelled and some trailed sugar beet harvesters have a front-mounted, full-width defoliator flail unit that removes most of the foliage before the roots are topped. There are two designs of flail defoliator, both driven by a hydraulic motor, and they remove most of the green leaf before the roots are topped. One type shreds the leaves and spreads them between the rows of sugar beet. The second type conveys the shredded leaves sideways on to cleared ground. Scalper blades, skew bar toppers or a feeler wheel topper unit complete the topping process.

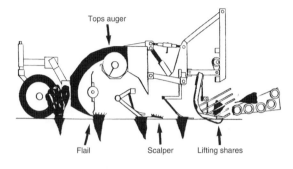

Figure 18.1   A flail rotor and scalper blade topper.

Plate 18.5   A 465 kW (625 hp) engine, three driven axles and a 30 tonne tank are included in the specification for this six-row sugar beet harvester. (Grimme)

Scalper blade toppers, which follow the defoliating flails, have sharp knives which slice through the top of the roots and remove any remaining leaf. This is then conveyed sideways by an auger and spread on cleared land ready to be ploughed in. Feeler arms automatically control the height of the scalper blades, which can move up and down on a parallel linkage or a spring suspension system. Either fixed or rotary tines sweep the remaining top debris off the topped rows.

Skew bar toppers, also used with a defoliator, consist of a small horizontal rotor with blunt-angled blades for each row, which rotate at a high speed. As the harvester moves along the rows, the rotors ride up and down over

Plate 18.6   A scalper blade topper unit with height sensors. (Garford)

Plate 18.7   A skew bar topper. (Garford)

the sugar beet, milling off remaining green material to leave a clean crown with a slightly domed top. The belt driven rotors float independently of each other and adjustable springs control the pressure exerted by the toppers on the sugar beet.

Feeler wheel toppers are chain driven and individually mounted on pivoting frames that allow them to float up and down over the rows. Springs support the weight of the feeler wheels, which lift and lower the topping knives as they run along the tops of the roots. Adjustment is provided to vary the distance between the bottom of the feeler wheel and the knife. A trash disc clears rubbish away from the topper arm, and rotating flails clear the tops away to leave the rows clear for the lifters.

Lifters

Sugar beet harvesters lift the crop with fixed or vibrating shares or squeeze wheel lifters. Many harvesters have an automatic guidance system to keep the lifters on the row and an automatic electro-hydraulic system to regulate the working depth of the lifters.

Shares are used to lift the roots on some harvesters. Fitted in pairs, the triangular-shaped shares squeeze the roots from the ground and pass them backward on to a rod link elevator or a cleaning turbine.

Vibrating shares, or oscillating shares, have an eccentric cam mechanism which oscillates the shares, loosening the roots as they move through the soil. The oscillating speed of the shares can be varied to suit soil conditions.

Squeeze wheel lifters consist of a pair of heavy spoked wheels with vee-shaped rims which run at an angle to each other in the soil. The wheels are arranged so that the rims run close together when they are in the ground and move apart at the top. The wheels grip the roots as they enter the soil, and further rotation pulls them from the ground. High-speed rotating rubber

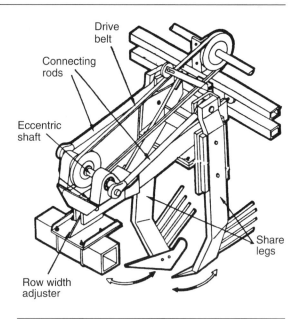

Figure 18.2   An oscillating share lifter.

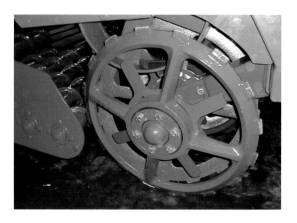

Plate 18.8   Squeeze wheel lifters. (Garford)

flails or cage wheels knock the roots from the lifting wheels on to the cleaning mechanism.

Cleaning

Turbines, rod link elevators and spiral rollers are the three main mechanisms used to clean the roots and elevate them directly into a trailer or into a large tank on the harvester.

Turbines The roots are passed from the lifters to a series of hydraulically driven rotary cleaning cages or turbines, which knock off the soil as they carry the roots towards the back of the harvester. Here, a rod link elevator conveys the cleaned roots to a holding tank on the machine. When full, a high speed side-elevator empties the tank into a trailer or onto a heap at the field headland at a rate of 10–15 tonnes per minute.

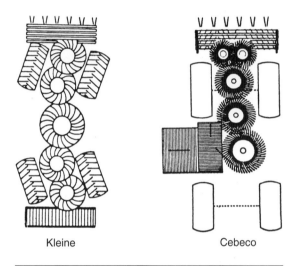

Kleine   Cebeco

Figure 18.3   The arrangement of turbine cleaners on a sugar beet harvester.

Plate 18.9   Sugar beet pass from the lifters onto these rollers, where they are pre-cleaned and then conveyed to the main cleaning area. (Garford)

Roller cleaners may consist of either open cage spiral rollers or solid rollers with spiral fins similar to those on an auger. The roots pass from the lifters on to contra-rotating rollers which scrub off the soil as the crop moves on to the turbines or a rod link elevator for further cleaning.

CONTROLS AND ADJUSTMENTS

Wheel Settings
Sugar beet are grown in various row widths, and most harvesters can work in row spacings of 450–600 mm (18–24 in.), but it is usual practice with self-propelled machines to drill the crop at a row spacing to suit the harvester. For trailed machines, the tractor wheels will need to be adjusted to match row spacing. A crop grown at 500 mm row spacings will need the tractor wheels to be set at 1.5 m.

The Topper
The topper units must be positioned to run along the middle of the rows of beet, and will need to be reset when changing to a crop sown at a different row width. This can be adjusted hydraulically on many harvesters, but spanners will be required on some machines. The roots should be topped through the crown. Under topping with leaf stalks left on the crown and over topping with the knife set too low should be avoided.

The height of the toppers on self-propelled harvesters is controlled automatically by sensors on the topper units that send signals to the electronic control system in the cab. Adjusting the spring pressure on the topper units varies the amount of top removed by scalpers and skew bar toppers.

The knife on a feeler wheel topper must be sharp. It has both vertical and horizontal adjustment to ensure the correct amount of leaf is removed and the top of the root is level. The tension on the springs, which allow the topper units to float up and down over roots of a different height, should be increased when the roots are small. In normal conditions, insufficient spring tension will result in the topper unit pushing the roots over before they are topped.

The flails, which sweep the severed tops off the rows, should be set to clear all green material away without knocking over any of the roots.

Lifters

Lifting depth is altered with the hydraulic rams which lift and lower the squeeze wheels or shares into and out of work. Most harvesters have automatic depth control with sensors linked to the in-cab computer. Older trailed harvesters may have manually adjusted depth wheels.

The lifters should be set deep enough to lift all of the roots without collecting any more soil than necessary. Excessive amounts of soil on the roots will reduce the efficiency of the cleaning mechanism, reduce output and give a dirty sample. If the lifting shares are not deep enough, some of the taproots will be broken off and left in the ground.

The distance between the bottom of the lifting wheel rims or the heels of each pair of shares can be adjusted to suit the size of the sugar beet. This adjustment is made from the cab or with a lever which moves the rims closer together or further apart or by changing the number of spacing washers between the wheels and their hubs. The lifter shares are bolted to legs, and the distance between the shares is increased when lifting large roots and reduced when the roots are small.

Oscillating share lifters have automatic depth control guided by sensors on the share arms, which also adjust the oscillating speed of the shares to suit soil conditions.

Cleaning

The roots will take more cleaning if the lifters are too deep. Harvesters with rod link elevators have agitators under them to help shake

Plate 18.10   The elevator on this trailed sugar beet harvester, which has a hydraulically driven axle, can empty the 12 tonne tank either when stationary or while harvesting the crop. (Garford)

the soil off the roots. Some rod link elevators have a stationary cleaning chain above the elevator that can be lowered to slow down the roots and knock off more soil.

The degree of cleaning achieved with turbine cleaners depends on their speed and the position of the adjustable gates which hold back the roots as they pass through the cleaning system. The turbines, driven by hydraulic motors, have a stepless variable speed control operated from the cab.

After cleaning, the roots pass on to a rod link conveyor, where more soil is removed as they are conveyed to the holding tank or side-elevated into a trailer. Tanker harvesters have a separate unloading elevator, which usually forms part of one side of the tank. The discharge elevator is driven by a hydraulic motor or vee-belt, and hydraulic rams move it outwards into position ready to empty the sugar beet on to the headland heap or into a trailer.

CLEANER LOADERS

Cleaner loaders are used to remove further soil from the sugar beet as they are loaded into the lorry taking them to the sugar factory. Cleaner loaders may either be high output self-propelled models or trailed machines used with a root bucket on a tractor front-end loader or a telescopic fork lift.

Self-propelled cleaner loaders are used on a field headland where the crop has been deposited by the harvester. This type of loader has a wide, spiral-based intake header to pick up and clean the heap of sugar beet, which may be up to 10 m wide. The cleaned roots are loaded onto a lorry by a high capacity elevator which is counterbalanced by a weight on the opposite side of the machine. This type of loader has an engine of about 225 kW (300 hp), and the

Plate 18.11   This self-propelled cleaner loader picks up the roots from the headland clamp and side-elevates them into a lorry or trailer. (Grimme)

Plate 18.12   Filled with a root bucket on a tractor loader or telehandler, this sugar beet cleaner loader fills a lorry in a matter of minutes. (CTM Harpley)

elevator, counterweight arm and intake fold down to 3 m for transport.

Trailed cleaner loaders consist of a large hopper on wheels with a cleaning mechanism and a rod link elevator driven by a diesel engine. A root bucket on a tractor loader or telescopic fork lift is used to load beet into the hopper, where spiral rollers or a rotary cage clean the roots, depositing soil, stones and trash under the machine. The roots receive further cleaning by a rod link elevator as they are loaded into a lorry at a rate of 2–4 tonnes per minute. Some older cleaner loaders have a picking table with space for two people to remove stones and trash. A self-propelled cleaner loader gradually moves forward as it loads, but a trailed cleaner loader remains static while loading the lorry, which is required to move forward as it is loaded.

POTATO PLANTING MACHINERY

A deep seedbed suitable for potatoes can be prepared on most soils with a plough and selection of cultivation machinery. However, growing potatoes on stony land can cause problems, including a high rate of wear and tear on harvesting machinery and a poor quality sample when the crop is harvested. Growing potatoes on a bed system, using a de-stoner to remove the stones and clods from the cultivated beds and depositing them in channels between the beds overcomes these problems.

The bed system of growing potatoes requires a deep seedbed, between 1.8 and 2.0 m wide, prepared with a bed former and a de-stoning machine. For all subsequent work, the tractor wheels are set wide enough to straddle the beds to avoid soil compaction, damaging the potatoes and making more clods in the bed.

Bed Formers

A bed former may either be a tractor-mounted rotary tiller with shaping boards or a toolbar with two or more large ridging bodies.

The power take-off driven rotary bed former (Plate 18.13) cultivates and de-stones unbroken soil and forms two 1.8 m wide beds for planting potatoes and other root crops in a single pass. The machine consists of a rotary cultivator followed by four rows of hydraulically driven de-stoning rotary stars, driven by a hydraulic motor, which bury the stones, and two bed formers which shape the beds.

The bed former (Plate 18.14), which can have two, three or four ridging bodies on a toolbar, is

Plate 18.13 Suitable for working on unbroken ground, this bed former cultivates the soil, buries the stones and shapes the ridges ready for planting potatoes or other root crops.

used to make wide ridges or beds in previously cultivated soil. Depending on the row spacing, the beds are usually between 1.5 and 2 m (60 and 80 in.) wide, and repositioning the ridging bodies on the toolbar changes the bed width. Subsoiler tines can be attached to the toolbar in front of the ridging bodies and markers are used to ensure accurate spacing between each group of beds.

De-Stoners

Destoners are used to lift beds previously made with ridging bodies, sieve out stones and clods and return the soil to the ground in shaped beds ready for the potato planter. The stones and clods are cross-elevated to one side of the machine and deposited into a furrow between two ridges. Depending on model, a de-stoner makes stone-free beds between 1.5 and 2 m wide.

Floating diablo rolls control share depth and keep the machine on the pre-formed ridges. De-stoning machines, usually with hydraulically steered rear wheels, have a combination of web-type conveyors and star-shaped agitator rollers.

The ridges are lifted by wide shares and pass along a web conveyor to rows of power-driven star-shaped agitators, which shake much of the

Plate 18.14 Potato beds being shaped with a three-row bed former before a de-stoning machine is used to remove stones and clods. (Grimme)

Plate 18.15 De-stoners reshape the pre-formed beds after separating out the stones and clods, which are cross-conveyed into a furrow between two potato beds. (Grimme)

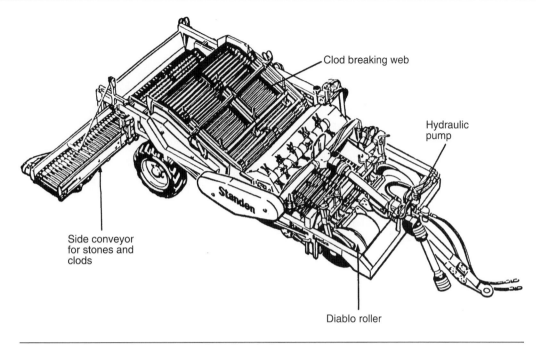

Clod breaking web

Hydraulic pump

Side conveyor for stones and clods

Diablo roller

Figure 18.4   De-stoner for a potato seedbed. (Standen)

soil back to the ground. De-stoners may have one or two web conveyors and a short or long run of star agitators. Machines for heavy or sticky land have more rows of agitators than those used on lighter soils. Stones, clods and the remaining soil fall from the agitators on to a rear-web conveyor with a clod-crushing web above it to help break up any stubborn clods.

When the stones and clods arrive at the back of the machine, they fall on to a cross-conveyor, which carries them sideways and drops them into the furrow left between two adjacent beds. The stones, which are pushed into the ground by the tractor wheels when the crop is planted, help to drain away water that collects between the beds. Web conveyors with a different spacing (pitch) between the rod links can be used on a de-stoner. A web with a 40 mm pitch will let fewer stones through with the soil than a web with a wider 60 mm pitch. The angle of the intake web or webs can be adjusted with hydraulic rams. A steeper angle will improve stone separation in difficult working conditions.

Potato Planters

With few exceptions, potato planters, which may be mounted or trailed, are fully automatic and plant between two and eight rows. Mounted potato planters used on de-stoned beds plant two, three, or four rows in one pass, and trailed machines plant four, six or eight rows. A hydraulic motor or land wheel may drive the planting mechanism. Planters used on pre-formed beds have a furrow opening coulter for each row. Either a belt or cup feed mechanism spaces the tubers in the soil, and shaping boards at the rear cover the tubers and reshape the beds.

Planters used to set potatoes on level fields have a furrow opening coulter for each row being planted and a feed mechanism which carries the tubers from the hopper and places them in the ground at regular spacings. Ridging bodies or angled concave discs cover the potatoes to complete the operation. Some planters have a ridge topping bar to leave flat-topped ridges.

Hopper capacities range from 500 kg on a two-row mounted planter up to 6 tonnes on an eight-row trailed planter. A fertiliser attachment for placing fertiliser near the potatoes, seed dressing attachments and granular applicators are among the optional equipment available for potato planters. Wide potato planters are folded hydraulically for transport and planters for chitted seed potatoes have frames on the toolbar to carry the chitting trays.

Cup Feed Planters

Automatic cup feed planters have land wheel- or hydraulic motor-driven belts or horizontal shafts with a series of cups which carry the seed potatoes from the hopper and drop them at regular spacings into the soil.

The automatic cup feed planter shown in Figure 18.5 has land wheel-driven endless belts with cups which collect single potatoes from the hopper and drop them at regular intervals into a furrow made by a coulter. A pair of concave discs – some planters have ridging bodies instead – cover the potatoes. Tuber spacing in the row is controlled with a gearbox which changes the speed of the cups. Some planters have a control unit in the cab to select the required spacing in the row.

Plate 18.16   This cup feed potato planter for unchitted seed consists of a power harrow to prepare the soil, furrow openers, planting mechanism and a ridging board at the back. (Grimme)

Another type of cup feed planter (Plate 18.16), also with furrow openers and large diameter discs or covering boards, has a horizontal shaft with the cups attached to discs spaced along the shaft. The cups pick up single potatoes from the hopper and drop them through tubes into the ground. The hopper has a hydraulic tipping system which, as the hopper empties, is used to maintain a full flow of potatoes to the cups.

Belt Feed Planters

Belt feed planters tend to space the potatoes less accurately in the rows, but they can be used at

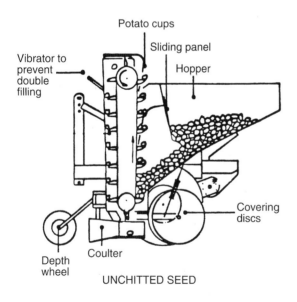

Figure 18.5   Cup feed potato planter for unchitted seed.

Plate 18.17   A four-row belt planter with a seed dressing attachment and a 3 tonne tipper hopper. (Grimme)

Plate 18.18   Belts in each hopper on this two-row belt feed potato planter convey potatoes into the lower chamber, where a shaped belt, driven by conical rollers, forms the potatoes into a single line and carries them to the outlet point, from where they fall into a shallow furrow made by a coulter. (Grimme)

high speed. Mainly used for planting unchitted seed, spacing in the row will be improved if the seed is graded for size. Many belt feed planters have a hydraulically tipped hopper to maintain the supply of seed potatoes to the seed belts. Some belt planters have a control unit in the tractor cab to regulate tuber spacing and monitor planter performance. The hopper is gradually tipped as the potatoes are moved on the mechanically or hydraulically driven and agitated main belt, and then passed on to a number of smaller shaped belts, one for each row being planted, which carry them to the ground immediately behind the furrow openers. Potatoes planted on ridged land are covered with shaping boards, and on level fields the crop is covered with ridging bodies or concave discs, which can be adjusted to leave a wide or narrow ridge.

Fertiliser Attachment

Fertiliser should not touch the potatoes when they are planted as it may damage the first stages of growth. For this reason, the fertiliser attachment on a potato planter places the fertiliser in one or more continuous bands close to the potatoes. The attachment, which consists of a hopper, a land wheel-driven metering

mechanism, flexible tubes and coulters, may be mounted at the front of the tractor or attached to the potato hopper. Fertiliser is metered into separate coulter tubes at a controlled rate and is placed in a band slightly below and to one side of the potatoes. Some fertiliser attachments have two coulters, which place fertiliser on both sides of the row.

Potato Planter Adjustments

Row width A row spacing of 750 mm (30 in.) is the standard setting on some planters, but on others the row spacing can be changed by moving the coulters on the toolbar to achieve the required row width.

Tuber spacing Belt feed machines tend to give a rather irregular spacing if the potatoes are not of an even size. Tuber spacings of 150 to 400 mm or more can be achieved by changing the speed of the belts. Tuber spacing in the rows is more accurate with cup feed planters. The cups are driven through a gearbox, and changing the speed of the cups alters the distance between the potatoes in the row. Spacing can be changed automatically on planters with an electronic control box in the tractor cab.

Planting depth Many potato planters have an automatic depth-control system monitored by a sensor wheel running on the field surface. This maintains the pre-set working depth of the furrow openers and the discs, ridging bodies or shaping boards used to cover the crop.

On other machines, the planting depth is controlled with depth wheels on the planter or by the tractor hydraulic system. The basic settings are made by raising or lowering the furrow openers and covering bodies in their mounting brackets. However, varying the depth of the covering bodies may alter the size and shape of the ridge.

Maintenance of Potato Planters

Regular lubrication of all moving parts and checking chain and belt tensions and tyre pressures are the main routine maintenance tasks. Fertiliser hoppers, especially those made from steel, should be emptied at the end of each day. Any moisture formed where fertiliser comes into contact with steel can quickly cause corrosion.

At the end of the season, bright metal parts should be protected from rust. Any parts which have been in contact with fertiliser should be thoroughly cleaned, washed and protected from corrosion.

POTATO HARVESTING MACHINERY

Potatoes may be lifted with a spinner or an elevator digger and left on the ground behind the machine for hand picking, but these relatively cheap potato lifters are only suitable for harvesting small areas of potatoes. Large-scale potato growers use either a tractor drawn or self-propelled complete harvester.

Spinners and Elevator Diggers

Potato spinners have a wide, flat share that runs under the potato ridge with either one or two tined spinner rotors that throw the contents of the ridge sideways against a net or canvas curtain, where most of the soil is separated from the crop. The potatoes fall to the ground for hand picking. Potato spinners are power take-off driven and mounted on the three-point linkage. Share depth may be controlled hydraulically or with a depth wheel.

Elevator diggers have a wide share running under the ridge which lifts the crop and passes it on to a rod link conveyor. Agitators under the conveyor shake the soil through the rod links as the potatoes are moved to the rear and fall back to the ground for hand picking. Minimum agitation is needed in light dry soil, but extra agitators may be required on wet or heavy ground. One- and two-row power take-off driven elevator diggers may be mounted or semi-mounted and have depth wheels to control the working depth of the share. A

slip clutch protects the digger from damage through overload or blockages.

Using Potato Diggers

Maincrop potatoes usually have the tops removed with a haulm pulveriser if the foliage has not died off. Early crops sold as new potatoes may be lifted with the tops intact.

Some farmers leave the headlands unplanted to make space for turning the diggers when lifting the crop. Others plant the headlands and leave an unplanted strip further into the field, again to make more turning space after the headland rows have been cleared.

The potatoes will be damaged if the share is not deep enough, and if the share is too deep, the machines will be overloaded and some of the potatoes will be buried.

Elevator diggers can be modified to lift other crops, including carrots, onions and flower bulbs. Specially shaped lifting shares and closer spaced rod link conveyors are required when harvesting some of these crops.

Haulm Pulverisers

Haulm pulverisers are sometimes used when the crop has to be harvested before the tops have died off, and removing the tops a few days before lifting makes it easier to harvest

Plate 18.19   A flail rotor on this two-row power take-off driven haulm topper removes the potato tops, and deflector plates guide them between the rows. (Grimme)

the crop. Tractor-mounted haulm pulverisers have steel flails across the full width of a high-speed rotor driven from the power take-off through a gearbox and multiple vee-belts, and a slip clutch to protect the drive from overload. The flails are shaped to match the contours of the potato ridges, and deflector plates guide the haulm between the rows. Some pulverisers, used with a complete harvester, have a cross-conveyor that moves the haulm sideways onto cleared ground. Haulm pulverisers may be front- or rear-mounted on a tractor or at the front of a self-propelled potato harvester.

Adjustable wheels or skids maintain a constant working height and ensure minimum damage to the ridge. A typical two-row pulveriser needs a tractor of at least 40 kW (55 hp), and the flails can be arranged to suit different row widths. Some haulm pulverisers can, after fitting a different set of flails, be used to clear tops from other root crops.

POTATO HARVESTERS

Potato harvesters will either elevate the harvested crop into a large self-emptying tank or bunker which is emptied into a trailer when it is full, or they may side-elevate the crop straight into a trailer pulled alongside. The lifting mechanism is very similar on most harvesters, but various methods of mechanical cleaning and sorting are used. Some smaller or older harvesters have a hand-sorting platform.

Other crops, including carrots, onions and parsnips, can be lifted with a potato harvester equipped with suitable lifting shares and elevator webs.

Trailed Harvesters

Trailed potato harvesters have a wide lifting share, or shares, which are guided under the rows of potatoes by two or more diablo rollers which run in front of the shares and on top of the ridges. Some harvesters have a hydraulic depth-control system with diablo rollers, which supports the rollers and at the same time reduces soil compaction and limits crop damage. The contents of the ridge are lifted on to a rubber-covered web or rod link elevator, and much of the soil is shaken through the links as the potatoes are carried up the elevator. Haulm extraction rollers at the top of the elevator separate out the potato tops and return them to the ground.

The potatoes, together with any remaining stones and soil, pass to the separator unit, where a number of star-shaped rollers or a

Plate 18.20   Hydraulic rams on this two-row trailed harvester control the discharge height of the side-delivery elevator to minimise the damage to the potatoes as they drop into the trailer. (Grimme)

Plate 18.21   A picking-off platform and a large bunker are features of this two-row harvester. (Grimme)

series of steel- or rubber-coated rollers sort the crop. Final cleaning and removal of remaining stones and small potatoes takes place as the crop moves forward along the separating rollers to a cross-conveyor and an elevator which carries the potatoes to the holding bunker or to the side-delivery elevator. To reduce tuber damage, a hydraulic ram is used to gradually raise the elevator as the trailer is filled, to minimise the distance the potatoes have to fall.

Trailed harvesters, especially those used on heavy or very stony soils, may have a platform with a picking-off table for two or more workers to remove clods and other rubbish before the potatoes are conveyed to the bunker or side-elevated into a trailer.

The main elevator and other working parts are usually power take-off driven, but hydraulic motors are also used on some harvesters. Hydraulics play an important role in the operation of a potato harvester. Hydraulic rams connected to the tractor hydraulic system and operated from the tractor cab control the working depth of the lifting shares and keep the machine level. Rams are also used to raise and lower the digging shares. Some harvesters

have hydraulic steering, which improves manoeuvrability on the headlands and keeps the lifting shares on the row, especially when working in difficult conditions.

Self-Propelled Harvesters

Large-scale potato growers may use a two- or four-row self-propelled harvester, either with a large bunker and a high capacity side-unloading conveyor or a side-elevator, sometimes with a holding bunker, for loading the harvested crop into a trailer. Some harvesters have a picking-off table, and a haulm pulveriser can be mounted on the front of the machine.

The two-row self-propelled harvester (Plate 18.22), with hydraulic motors driving the main components, has a front-mounted haulm pulveriser and lifting shares with diablo roller depth control. The potatoes, soil and stones are carried up the intake web to the haulm separation mechanism and then to the separator rollers, where small potatoes along with any remaining stones and soil are removed. The crop is then elevated up to the picking table where, if required, the potatoes can be sorted by hand before being elevated into the bunker,

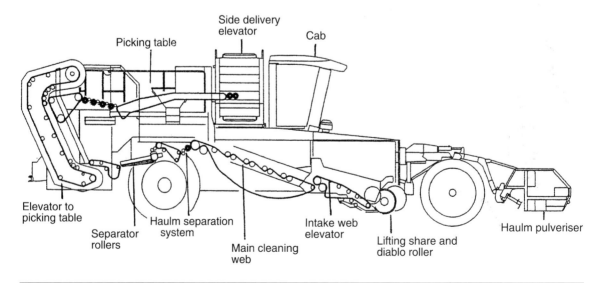

Figure 18.6   Crop flow through a two-row self-propelled potato harvester. (Grimme)

Plate 18.22   This self-propelled two-row potato harvester, with a 209 kW (280 hp) engine, has a front-mounted haulm pulveriser and a 7 tonne bunker. (Grimme)

which if necessary can be emptied into a trailer pulled alongside while harvesting continues.

Depending on size, self-propelled potato harvesters have an engine in the 190 to 372 kW (250 to 500 hp) bracket and a 7 to 15 tonne capacity bunker with a high output discharge elevator. Hydraulic motors are used to drive all main components, and on some models there is a choice of wheels or rubber tracks at the rear to spread the weight of the harvester. Cab equipment includes joystick controls, and some harvesters have a display screen linked to video cameras to give the driver a complete view on the critical working areas of the machine.

CHAPTER 19

Mechanical Handling

Fertiliser, seed and other materials are usually delivered to the farm in bulk, either in half-tonne or one-tonne bags, and sometimes on pallets. Most farms have a front-end loader on a tractor, a telescopic handler or a mast-type fork lift truck to handle pallets and big bales. They are also used to load lorries with grain leaving the farm using a large bucket, to handle farmyard manure or to fill or empty silage clamps with a large fork.

TRACTOR FRONT-END LOADERS

A front-end loader has two heavy-duty arms, known as the boom, which pivot on a frame attached between the front and rear wheels of the tractor. Oil from the tractor hydraulic system is pumped to rams attached to each side of the boom to raise the loader. Most loaders are lowered by their own weight, which forces oil from the rams back into the tractor when

Plate 19.1   The hydraulically operated grapple on this front-end loader enables the driver to pick up full loads of manure. (John Deere)

the driver uses the control lever to lower the loader.

There is a wide range of attachments for front-end loaders. They include a manure fork or grab, grain bucket, silage grab, bale handling equipment and a pallet fork. Auxiliary rams are used for various tasks, including tipping a manure fork or grain bucket and gripping big bales. These rams are supplied with oil from the hydraulic spool valves at the back of the tractor. Some attachments have a hydraulic or mechanical self-levelling linkage to keep the bucket, big bale or pallet parallel with the ground while the loader is being raised. The maximum lifting height and capacity depends on the size of the tractor power and type of loader. A typical front-end loader on a 40 kW (55 hp) tractor has a maximum lift of about 1 tonne to a height of about 3 m, and a loader on a 90 kW (120 hp) tractor will lift 2 tonnes or more to a height of about 4 m. Front-end loaders can be difficult to fit to older tractors, but most modern loaders have plug-in hydraulic connections and quick-attach brackets secured with pins. They

are equally easy to detach and have their own parking stands.

Using Front-End Loaders

Loaders put a considerable strain on the front wheels and axle of a two-wheel drive tractor. The tyres should be suitable for heavy work, and inflation pressures need to be increased by about 1 bar (15 psi) above front tyre pressure for fieldwork.

A counterbalance weight, usually a heavy cast iron block, attached to the rear hydraulic linkage, improves tractor stability and reduces wear on the steering linkages. Power-steered four-wheel drive tractors with a forward/reverse gear shuttle have overcome the problems of heavy steering and worn clutches encountered with two-wheel drive tractors used extensively for loader work.

The height of the centre of gravity of a tractor will be raised when a fully laden front-end loader is raised to its maximum height. A high centre of gravity reduces the stability of a tractor and increases the risk of overturning, especially when cornering at a fast speed. For safety

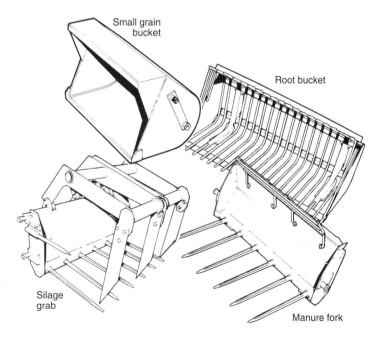

Figure 19.1   Forks and buckets for telescopic fork lifts and tractor front-end loaders.

Plate 19.2   This front-end loader has a big bale spike attachment. (John Deere)

Plate 19.3   Loading silage with a telescopic fork lift which is also equipped with a trailer hitch. (Massey Ferguson)

purposes, when travelling, the loader should be raised high enough for the driver to have full vision in front of the tractor, but not raised to its full height.

SELF-PROPELLED MATERIALS HANDLERS

Telescopic handlers, mast-type forklifts and skid-steer loaders are used to move bulk materials around the farm. Originally made for lifting and stacking pallets, forklifts became popular in the 1980s for handling farmyard manure, grain, straw, sugar beet, fertiliser and other materials. Industrial fork trucks with small wheels and a petrol, diesel or electric power unit are only suitable for use on concrete and other hard surfaces. Electric- and gas-powered models are sometimes used in farm pack houses and vegetable stores.

Telescopic Handlers

Telescopic handlers have almost replaced the earlier mast-type forklift on many farms. Rigid chassis and articulated telescopic handlers have a two- or three-section extendable boom on the right-hand side of the driving cab on a rigid machine, and in front of the cab on an articulated model. Boom attachments include a manure fork, bale and silage grabs, pallet fork and a grain bucket.

Telescopic fork lifts with engines in the 50–75 kW (65–100 hp) bracket with a powershift, hydrostatic or CVT transmission have auxiliary hydraulic services to operate a manure fork, grain bucket and other attachments. Smaller models have a maximum lifting capacity of between 1.5 and 4 tonnes, to a height of between 4 and 5 m. Larger machines with a 75–112 kW (100–150 hp) engine will lift up to 7 tonnes or more to a height of between 7 and 9 m. Many rigid chassis telescopic handlers have four-wheel drive and four equal-sized wheels, which can be steered either by the front wheels for road work, or by all four wheels for tight turns in the field or farmyard. Some machines also have crab steering (see page 74 in Chapter 6), which allows the handler to move forwards and sideways at the same time – a useful feature when, for example, manoeuvring in narrow feed passages. The steering mode is selected with a switch in the driving cab. Many telescopic forklifts have a top speed of 40 kph (25 mph), but only in front-wheel steering mode.

Articulated models have similar engine, lift height and lift capacities to rigid chassis models of equivalent sizes, but with centre-pivot steering and a reduced turning circle, they are more manoeuvrable. A typical articulated handler with an 88 kW (120 hp) engine and

Plate 19.4   Handling farmyard manure with a telescopic fork lift. (Claas)

hydrostatic transmission will lift 3 tonnes to a maximum height of 7 m.

Multi-function telehandlers can be used to move bulk materials, or for field work with implements either hitched to the drawbar or three-point linkage. They have an engine in the 75–90 kW (100–120 hp) bracket, hydrostatic transmission and four-wheel drive with front wheel steering for roadwork. Four-wheel steering provides a smaller turning circle for handling materials and fieldwork.

Plate 19.5   This dual-purpose telescopic fork lift with hydrostatic transmission, hydraulic three-point linkage and power take-off can be used for handling or for field work. (Merlo)

Mast–Type Forklifts

This type of forklift, which has been replaced with a telescopic handler on many farms, is rear-wheel steered with a mechanical or hydrostatic transmission, and has a two-stage front lifting mast and either two- or four-wheel drive. Mast-type forklifts, which have limitations in reach and lifting height, are used mainly for handling boxes and pallets. The mast can be tilted forward to improve reach when loading a trailer. As the load is carried directly in front of the driver, the forks must be lifted quite high when carrying a bulky load for the user to see where he or she is going. However, for maximum stability, the load should be carried as low as possible without obscuring the driver's view.

SAFE USE OF MECHANICAL HANDLING EQUIPMENT

Users of all types of materials handlers must be trained in their use and hold a relevant Certificate of Competence.

It is important to remember that as the telescopic boom is extended, the maximum safe load the machine can lift and still remain stable will be reduced. A safe load indicator in the driver's safety cab warns the driver when the loader arm reaches its safe loading height and forward reach. A built-in safety mechanism that stops the boom extending too far forward to avoid the risk of the machine tipping forward when lifting a heavy load is fitted to all modern machines.

Because the centre of gravity of a telescopic handler rises as the load is lifted, the boom should be kept as low as possible when driving with a loaded fork, bucket or big bale grab. Particular care is needed when driving across a slope with a loaded machine.

Skid–Steer Loaders

These versatile machines are suitable for working in very confined spaces – clearing manure from low livestock buildings is a typical application. A skid-steer loader has a separate transmission system for the two wheels on each side of the machine, with a chain and sprocket drive, a hydraulic motor or hydrostatic transmission.

Skid-steer loaders are steered with hand levers, which lock, or partly lock, the wheels on one side while full drive is maintained to the wheels on the opposite side, so that the loader can almost turn within its own length. This type of loader varies in size, from a 15 kW (20 hp) machine with a lift capacity of 350 kg, to much larger models which can handle a load of about 1 tonne.

Farm Transport

Many types of material have to be transported around the farm, and from time to time workers need to travel from place to place as quickly as possible.

ALL–TERRAIN VEHICLES

ATVs, as they are commonly known, can negotiate most ground conditions and are used by many farmers as a means of getting round the farm. As well as providing personal transport, two- and four-wheel drive ATVs are used for light work on arable crops including fertiliser spreading, applying slug pellets and spraying small areas where pests or weeds are a problem. With their low ground pressure of about 0.15 bar (2.5 psi), ATVs can be used to do this work when the soil is too wet to use a tractor. Some livestock farmers use an ATV to help with the management of stock, especially on outlying fields. In this way they save a lot of walking, and can also be used with a small trailer to carry fodder, fence posts, etc.

ATVs have a single-cylinder, air-cooled four-stroke petrol engine or a single-cylinder, water-cooled, diesel power unit. Earlier petrol-engined models have a carburettor fuel system,

but many of the more recent machines have a fuel injection system. The more powerful engines usually have an oil cooler.

Cylinder displacement, or capacity in cubic centimetres (cc), is usually given in the specifications for ATV engines instead of kW or hp. The engines with a cylinder capacity of between 300 and 700 cc run at speeds of up to 7,000 rpm, and power output is typically 9–27 kW (12–35 hp).

Electric starting is standard on all but the most basic models. Engine speed is controlled with a throttle lever on the handlebars, and an automatic centrifugal clutch engages drive to the gearbox and either a shaft or chain puts drive to the wheels. ATVs may have three, four or five forward gears, and one or two in reverse, with manual or electronic gear selection. Others have an automatic variable speed transmission with a high/low and reverse range lever. Power steering is standard on the more powerful models, which may also have a rear suspension system with shock absorbers and a diff-lock. Depending on model, an ATV may have shoe or disc brakes or a combination of both braking systems.

Plate 19.6   Many ATVs have a single-cylinder four-stroke engine, selectable two- or four-wheel drive and power steering. (Suzuki)

A 12-volt electric socket provides the power to operate the electric motor on a broadcaster, crop sprayer or a slug pellet applicator either mounted on the ATV's load platform or towed behind the machine. A separate single-cylinder air-cooled engine is used to drive larger implements, including mowers, sprayers and fertiliser broadcasters hitched to the drawbar.

Safe Use of ATVs

Lack of care when driving an ATV on slopes and very uneven surfaces can lead to accidents, especially when travelling at excessively high speeds.

- All first-time users of an ATV should attend a training course. The dealer selling the machine can usually arrange this.
- Always wear a helmet and protective clothing when driving an ATV.
- Drive at a speed suitable for the ground conditions.
- Don't make sharp turns at high-speed – this may cause the machine to roll over sideways.
- Always drive directly up or down steep slopes and take great care when turning.

TRAILERS

Every farm has a range of different trailers. Most of them are of steel construction, but older trailers with wooden sides and floors and a jaw hitch for a tractor drawbar are still in use. Steel trailers may have a single or tandem axle, and some high capacity models have a triple axle. Most trailers have a drawbar with a ring for a pick-up hitch, but some have a combined jaw and pick-up hitch arrangement.

There are many different types and sizes of farm trailer, including tippers, side tippers, flat

Plate 19.8   A tipping trailer with twin rams and a tandem axle.

Plate 19.7   Cutting grass with an ATV and a flail mower driven by a small petrol engine. (Suzuki)

Plate 19.9   This high-sided silage trailer has twin rams and a triple axle. (Joskin)

Plate 19.10   Grain being transferred to a chaser bin while harvesting continues. The bin's auger allows rapid unloading into trailers or lorries on the edge of the field for transport to store. (Massey Ferguson)

beds, drop sides, dumpers and livestock trailers. There are also specialist trailers, and conversion kits for standard models, for carting grain, silage, bales, etc. Tailgates on modern high-sided trailers used for grain, sugar beet and silage making, etc are opened and closed with hydraulic rams. The load capacity of farm trailers ranges from 1 tonne for compact tractors up to 16 tonnes or more.

Trailers for special purposes include chaser bins with an emptying auger, driven by a hydraulic motor, used in the harvest field to collect grain from a combine harvester and empty it into a waiting trailer or lorry. Bulk handlers, some with built-in auger, are used, for example, to fill a fertiliser spreader or grain drill hopper.

The law requires that trailers with a jaw-type hitch must have a screw jack to raise and lower the drawbar. There are also regulations concerning the provision of a braking system when a trailer is used on the highway. Although some trailers only have a parking

Plate 19.11   Moving cattle with a livestock trailer. (Joskin)

brake, trailers designed for tractors with a top speed of 40 kph (25 mph) have hydraulic brakes. They may be linked to the tractor hydraulic system, but trailers over a certain laden weight have their brakes connected to the tractor braking system. Trailers towed by tractors with even higher road speeds are required to have air-operated brakes connected to the tractor braking system. Some trailers have dual braking, with separate hydraulic and air-operated cam expanders on the brake shoes.

Trailers used on the public highway must have a full set of working road lights. They may either be installed permanently on the trailer, or on a lighting board attached to the back of the trailer or other trailed implement. Farm trailers and detachable lighting boards, attached to the back of the trailer and connected to the tractor lighting system, should have rear and stop lights, direction indicators, reflectors and a number plate light. When driving on dual carriageways, amber warning beacons must be visible from behind the trailer. Highway marker boards must be used to warn other road users when towing a wide or overhanging load.

CHAPTER 20

Farmyard and Estate Machinery

Farm machinery includes much more than the equipment used for working in the fields. Farm and estate machinery covers a wide range of equipment from that used to dry and handle grain to machines used for hedging and ditching. Farms with livestock will have animal feeding equipment, and milking equipment will be in daily use on dairy farms.

GRAIN DRYING AND STORAGE

Freshly harvested grain may have a moisture content anywhere between 12 and 25 per cent. During most harvests, grain will come off the combine at 14–18 per cent moisture content. For safe, long period bulk storage, the moisture content of wheat, barley and oats should not be more than 15 per cent.

Plate 20.1   A continuous flow high temperature drier incorporating a diesel or kerosene-fired heater with an output of 20 tonnes/hour when drying wheat of 20 per cent moisture down to 15 per cent. (Bennett & Co)

Many farmers have sufficient combine capacity to harvest more grain than they actually grow. This surplus capacity makes it possible, in most years, to delay harvesting until the grain is dry enough to store direct from the combine. Some drying will be necessary during difficult harvesting seasons, but a high capacity combine harvester will help to reduce the amount of grain which has to be dried.

Grain with a high moisture content will heat and, before long, moulds will appear, making the grain unsaleable. Insects and mites, which may still be present in the store and will eat into the grain even if the store was thoroughly cleaned before harvest, can only work and multiply when there is sufficient oxygen and moisture and a high enough temperature. To safely store grain for long periods, it should be placed at no more than 15 per cent moisture into a clean store to prevent heating and/or damage by insect pests.

Drying the grain with a high temperature drier or a warmed air drier to reduce the moisture content is the usual method of preparing grain for long-term storage. However, an alternative method of storing damp grain in airtight silos to exclude oxygen, in the same way as grass is stored in airtight silos or plastic bags, has been used on some farms where the grain is to be used for feeding livestock.

Grain Driers

There are three main types of grain drier. One type uses relatively high temperature air to remove the moisture quickly and then cools the grain before it is stored. Low temperature driers use air heated to a few degrees above ambient temperature (the actual air temperature) and drying takes place over several days. Low temperature driers may either be permanent installations in buildings or portable driers on wheels that can be towed to various locations around the farm.

High Temperature Driers

The most common type of drier consists of a galvanised steel vertical tower with perforated sides, two fans and a heater. It is usually oil-fired, but some low output driers have an LPG gas-fired heater. After damp grain from the combine has been cleaned, it is elevated to the top of the drying tower. The grain then falls very slowly at a controlled rate while heated air is blown through the perforated sides of the tower, to reduce its moisture content. A second fan in the lower section of the tower blows unheated air through the grain to cool it. Once dried and cooled, the grain is stored in bins or in bulk on the floor of the grain store.

Another type of high temperature drier has a continuous horizontal conveyor for the grain, with fans and a heater below. Hot air and then cold air are blown through the bed of grain as it moves through the drier. Horizontal driers require a much larger floor area than vertical driers.

It is important to control the maximum temperature of the air used in a high temperature grain drier. The safe air temperature when drying any grain intended for seed or malting barley with a harvested moisture content below 24 per cent must not exceed 49° C. This is reduced to 43° C for grain above 24 per cent moisture content when harvested. Temperatures above this level will impair germination. The maximum safe temperature for milling wheat and barley harvested at below 25 per cent moisture content is 60° C, and for grain above 25 per cent moisture content it is 60° C. Higher temperatures can be used when drying grain for stock feed.

The output of high temperature driers is measured in tonnes per hour, at a moisture extraction rate of 5 per cent when drying wheat. Output also depends on the temperature and relative humidity of the drying air.

A typical range of continuous flow high temperature driers has outputs ranging from 10–40 tonnes an hour when drying malting

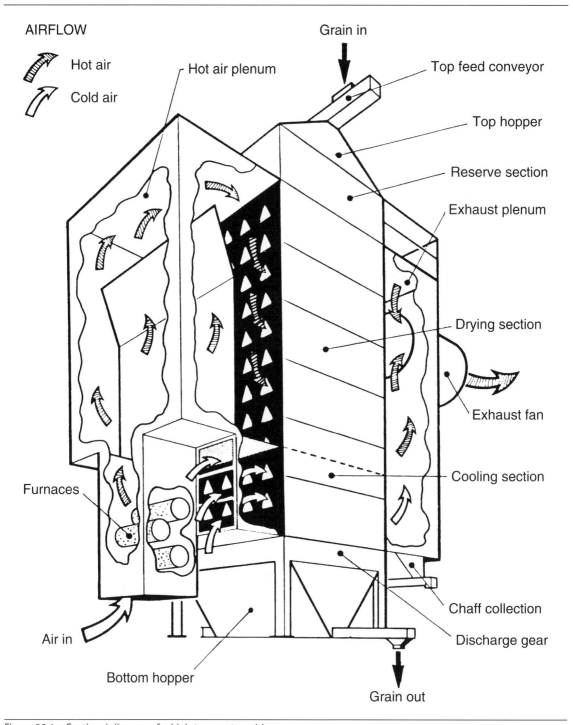

Figure 20.1   Sectional diagram of a high temperature drier.

Plate 20.2   These 1,000 tonne capacity grain bins have an 18 tonne capacity intake pit, a grain stirring system and an oil-fired burner with a large fan for drying the grain. Each silo has an unloading auger used to fill a lorry at a rate of 100 tonnes/hour.

barley, and from 15–50 tonnes per hour when drying milling wheat.

Low Temperature Driers

Ventilated bins and grain storage buildings with under-floor ventilation and portable driers are the main types of low temperature grain drier. Bins and storage buildings use slightly warmed air blown upwards through a perforated floor to dry the grain. Drying takes place slowly, with the moisture content reduced by about 1 per cent per day. Many low temperature driers have a fuel oil or LPG gas heater, but biomass heaters, which burn straw, are used on some farms. The depth of the dried grain in the store gradually increases upwards until the complete bin or floor store is dry. The moisture content may be increased if air is blown through the grain on a damp or misty day.

Although no longer in common use, some ventilated bins have a central perforated air tube through which warmed air is blown radially outwards through the grain to the perforated outer walls. A large plug in the centre tube can be used to control the height of the drying zone in the bin. Another type of ventilated bin has a perforated floor above an air chamber and the grain is dried by blowing warm air upwards through the contents of the bin.

Most ventilated floor stores have a perforated floor above a series of ducts connected to a large fan which blows warmed air upwards through the grain. This design leaves the storage floor area completely clear for a tractor or telescopic loader with a grain bucket or a wheeled sweep auger to work in the store.

An alternative ventilated floor storage system has a series of floor level ducts connected to a

Plate 20.3   The sweep auger on the floor of this grain bin carries the grain to a central hopper, from where a horizontal auger takes it to the outside of the silo at a rate of about 30 tonnes/hour. The vertical auger is used to stir the grain stored in this 200 tonne capacity silo. (Bennett & Co)

Plate 20.4   A ventilated floor grain store which is filled by a travelling overhead conveyor system. (Bennett & Co)

Plate 20.5   These low volume fans with 7.5 kW motors are used to ventilate previously dried grain. (Bennett & Co)

main tunnel with a heater unit and fan at one end of the tunnel. Warmed air, blown through the grain from the ducts, is used to dry the grain. Blanking plates are used to close off the air supply from the tunnel to the ends of each section of the floor ducts and control which part of the drying floor receives warmed air. It is important to have the grain at an even depth over the drying floor to give equal resistance to airflow and ensure even drying throughout the store.

The store is usually filled by reversing trailer loads of grain into it and tipping them onto the heap. The grain is then pushed up and levelled, usually with a grain bucket on a telescopic forklift or a tractor front-end loader. Some grain stores, including those with above-floor drying ducts or ventilated bins, have a holding pit for the grain brought to the store by trailers. Vertical and overhead conveyors carry the grain from the pit and fill the required part of the drying floor.

In some stores, the grain is put into a holding pit before being passed through a pre-cleaner to remove chaff and green material, but in good harvesting conditions the grain can usually be dried straight from the combine. A pre-cleaner has reciprocating sieves and a fan driven by electric motors which remove unwanted matter from the grain. If pockets of green material are allowed to build up in the heap, the grain may not dry evenly and hot spots may occur.

Plate 20.6   A section of a wooden floor in a ventilated floor grain store and the tunnels under it which carry the warmed drying air.

Plate 20.7   Driven either by a tractor power take-off or a three-phase electric motor, this portable 12 tonne capacity batch drier with a diesel-fired heater unit has an output of up to 6 tonnes/hour when drying grain down from 21 to 16 per cent moisture content. (Opico)

The store is usually emptied with a grain bucket on a telescopic forklift or loader, or with a sweep auger. High-speed loading of floor-stored grain is possible with a grain bucket on a telescopic handler. These loaders also have the advantage of a high reach for loading high-sided lorries. A sweep auger consists of an auger with left- and right-handed flights in an open-fronted trough which is pushed into the heap of grain. This auger feeds the grain to a high capacity elevating auger which loads the lorry or trailer.

Portable Driers

Fully portable recirculating grain driers can be towed with a tractor to remote storage buildings or used as a backup for a permanent grain drying installation. Some portable driers are completely self-contained and require no electricity. A combination of tractor power take-off and a gas or diesel engine provides the necessary services to operate the drier. Depending on size, portable driers have an hourly output of between 4 and 15 tonnes of dried grain.

There are two types of portable drier. One dries grain in batches; the other is a continuous

flow machine. Most portable batch driers have a cylindrical-shaped hopper with perforated walls with a thickness of approximately 450 mm of grain between the inner and outer walls. When filled, the grain in a batch drier is circulated many times by an auger in a central tube which lifts it from the bottom to the top of the hopper. A large fan driven by a diesel engine, tractor power take-off shaft or a three-phase electric motor blows warmed air into a vertical air chamber, about 300 mm in diameter, in the centre of the drier. From here the air, warmed by a fuel oil or an LPG gas heater, passes radially outwards through the perforated walls of the drier, and the moisture content is gradually reduced as the grain is circulated through the hopper. When the grain has been dried, the heater is turned off and air at ambient temperature is blown through to cool the grain before it is discharged into a trailer and taken to the grain store. Alternatively, a portable drier can be situated in a position where the dried grain can be discharged straight into a store.

Portable batch driers usually have an output of between 4 and 15 tonnes an hour or more, depending on the size of the drier and moisture content of the harvested grain. Hopper capacity varies from 8–40 tonnes. A typical 9 tonne batch drier with an output of 4 tonnes an hour when drying grain down from 20 to 15 per cent moisture requires a tractor of at least 30 kW (40 hp) at the power take-off to drive the auger and fan.

Portable continuous flow driers work in a similar way, but the grain is dried and cooled as it passes through the machine, and the dried grain is discharged from the bottom of the drier. Continuous portable flow grain driers have hopper capacities of 4–30 tonnes. A typical 18 tonne continuous flow portable drier has an output of up to 7 tonnes of dried grain per hour.

GRAIN CONVEYORS

Mechanical and pneumatic grain conveyors, which carry grain vertically or horizontally, should have sufficient capacity to empty the intake pit before the next trailer load arrives at the store. Grain conveyors must also keep the drier supplied with enough grain to keep it working at full capacity and to carry the dried grain to the store. The average combine harvester output is about 30 to 50 tonnes per hour in a standing crop of wheat and the capacity of continuous flow driers is in the region of 20 to 50 tonnes per hour or more. Larger combines and driers will, of course, exceed these output figures.

Most grain stores have a remote control system to start and stop the conveyors and micro switches to protect them from damage through blockages or overloading. A safety switching system is usually provided which stops all of the conveyors if one of them becomes blocked or develops a fault.

Belt and bucket elevators convey the grain vertically. They have an endless belt, usually made from canvas or rubber, with small closely spaced steel or plastic buckets which pick up grain from a hopper at the bottom of the elevator. The grain is carried to the top where it may be directed into a pre-cleaner, top loading drier or an overhead conveyor. Belt and bucket elevators in farm grain stores have a capacity of 20–100 tonnes per hour or more depending on the size of the cups and the speed of the elevators.

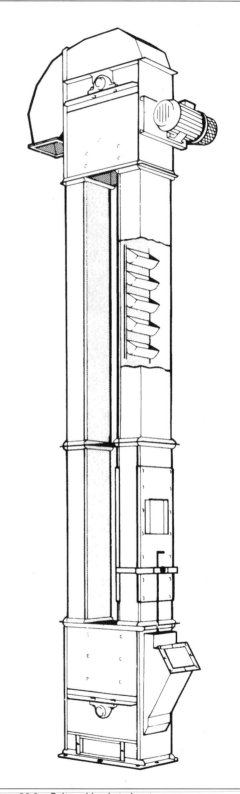

Figure 20.2   Belt and bucket elevator.

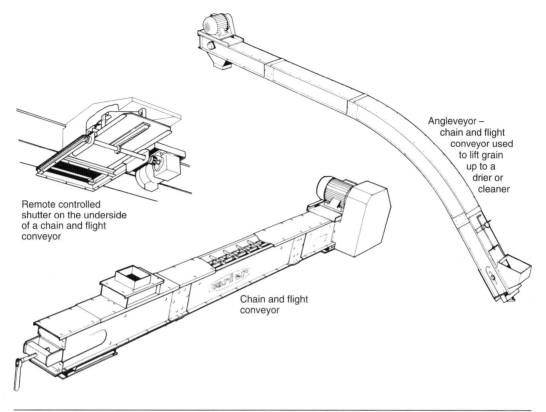

Angleveyor –
chain and flight
conveyor used
to lift grain
up to a
drier or
cleaner

Remote controlled
shutter on the underside
of a chain and flight
conveyor

Chain and flight
conveyor

Figure 20.3   Chain and flight conveyor.

Chain and flight conveyors are used for horizontal overhead conveying in the grain store. An endless roller chain with closely spaced heavy-duty plastic or steel flights moves the grain along the length of the conveyor trough. Discharge outlets, spaced at intervals along the length of conveyor trough, are used to fill the required section of the drying floor. The outlets are opened and closed either manually, with a system of cords, or with remote-controlled electric motors. Depending on size, chain and flight conveyors carry grain at a rate of 20–100 tonnes per hour or more.

Belt conveyors also carry the grain horizontally, and depending on size, they have a carrying capacity of 45–100 tonnes per hour or more. Belt conveyors are smooth running and are suited to high level installation, especially when conveying grain over some distance. The grain is carried on an endless high-speed

V-shaped rubber belt, which varies in width from 500–650 mm (20–25 in.).

Pneumatic conveyors have the advantage of conveying grain horizontally and vertically and

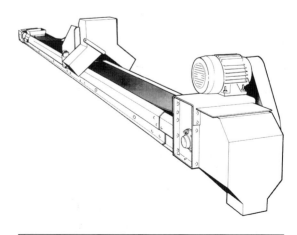

Figure 20.4   Belt conveyor.

around corners within the same system of 150 mm (6 in.) diameter ducting. A pneumatic grain conveyor consists of an intake hopper, a high capacity fan driven by an electric motor, a system of ducting or pipework and outlet points. There are suction and blower types of pneumatic grain conveyor. Both use a fan to create a high-speed airflow which carries grain, metered into the ducting from an intake hopper, to the outlet, which may be up to 100 m from the intake hopper. A typical farm pneumatic conveying system with a 40 kW (30 hp) fan has an output of up to 30 tonnes per hour. As a general rule, a 750 watt (1 hp) motor will deliver 1 tonne per hour in a blower system, but only half of this rate with a suction system.

Grain augers have the advantage of being portable and can be used to convey grain at any angle between horizontal and vertical. They have an adjustable discharge height, and most portable augers convey grain from ground level to the top of a heap of grain at an angle of 45 degrees or more. Augers are made in various lengths, with the smaller ones being placed in position by hand and larger high capacity models, some with an intake hopper at the lower end, being carried on a wheeled chassis. Augers used in farm grain stores are vee-belt driven by an electric motor and have an auger tube diameter of 100, 125 or 150 mm (4, 5 or 6 in.). The base of the auger has a hopper with a wire cage that covers it for safety reasons but allows the grain through; it is placed in a heap or has grain directly poured into it from an outlet in the trailer tailgate. The auger flights carry the grain to a discharge spout at the top of the tube, and in this way grain can be heaped high to maximise the capacity of the store. Where this type of auger is not used in an on-floor store, the grain must be pushed up with a grain bucket on a telescopic forklift.

Augers are mainly used for heaping up grain in the grain store, filling grain silos and emptying grain bins. The capacity of a grain auger depends on its diameter and working angle. The output from a typical 150 mm diameter auger when working horizontally is about 40 tonnes per hour, but when used in an almost vertical position, the output falls to 14 tonnes per hour.

Wheeled sweep augers are used to collect and load grain in some floor stores. A horizontal auger at the front of the machine picks up grain from the floor and passes it to a central elevating auger, which loads it into a waiting lorry or trailer. Circular rotating sweep augers (Plate 20.3) are used to empty some round grain bins.

Plate 20.8 A portable grain auger can be used to move grain over a short distance to fill or empty a floor storage system or to load a lorry from the store. (Astwell Augers)

MILLS AND MIXERS

Bulk loads of ready-mixed animal feed are delivered to many livestock farms, but some produce all or part of their requirements with a mill and food mixer, using home grown or bought-in feed grain to prepare their own livestock feed.

Roller Mills

These are used to crush or flatten grain to make it more digestible for farm livestock, the main example being rolled barley used to fatten beef cattle. A roller mill consists of two smooth steel rollers held slightly apart by spring pressure. The degree of crushing depends on the clearance between the rolls, and adjustment is provided to move the rolls closer together against the spring pressure. The rate of flow of grain from the hopper to the rolls is controlled with an adjustable slide. Stones and other debris in the grain, which could damage the rolls, are caught by a sieve in the hopper. After the grain has been rolled, it is either discharged and collected under the mill or is conveyed direct to the feeding area.

The rolls will wear very quickly if they are allowed to run against each other without any grain passing through them. Very dry grain below 17 per cent moisture content is not in the best condition for rolling. The product will be rather dusty and there will be a tendency for the grain to shatter.

Roller mills are suitable for automatic operation. A flap switch or a diaphragm switch in the hopper will turn off the power when the incoming grain supply is exhausted.

Hammer Mills

Hammer mills vary considerably in size and output. The most common hammer mills have a power requirement of about 3.75 kW (5 hp), but much larger models are in use. Most hammer mills are driven by a vee-belt from an electric motor. Hammer mills are usually installed under a large holding hopper and will grind a batch of grain without supervision. An

Plate 20.9   A small petrol-driven automatic hammer mill. Grain tipped into the hopper is sucked into the milling chamber. When the meal is fine enough, it passes through the rotary screen and is blown to the cyclone, where the air is separated from the meal before it is bagged off.

automatic cut-out switch in the hopper, which turns off the power supply when it is empty, allows the use of cheap rate electricity.

A typical hammer mill has a shaft, running at speeds of up to 6,000 rpm, with a grinding rotor at one end and a fan at the other. It is usual to have eight flails (hammers) attached to the rotor, with their tips passing very close to a fixed circular screen with a large number of small round holes. Grain is tipped into a low level hopper where suction from the fan draws the grain into the grinding chamber. The flails beat the grain against the screen until the resulting meal is small enough to pass through the holes in the screen. As well as drawing grain from the hopper into the grinding chamber,

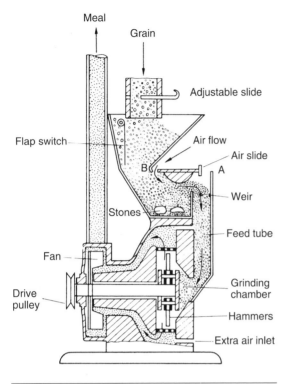

Figure 20.5   A hammer mill. Grain is fed from a hopper above this small automatic machine and the result is blown into a storage bin.

the fan also blows the meal, which has passed through the screen, to a mixer or storage bin.

Adjustments

Degree of grinding A hammer mill has a set of screens with different-sized holes, and the fineness of the meal depends on the size of screen used. A typical mill may have screens with 1.5, 3, 6 and 9 mm holes, and removing the grinding chamber cover gives easy access when changing the screen.

Feed rate The fan sucks air through the grain in the hopper and the strength of the airflow, controlled by an adjustable slide, determines how much grain is drawn into the grinding chamber.

Maintenance

Regular lubrication is essential. High melting point grease is recommended for the high-speed rotor shaft bearings.

The vee-belt drive must be at the correct tension to ensure full power is transmitted to the mill.

The striking face of the hammers should have square corners, and many mills have four square corners on each hammer. When the square corners on the hammers become rounded and worn, they can be reversed and also turned from end to end so that all four corners can be used. The rotor must be balanced to prevent vibration. If it is necessary to fit a new hammer, another must be fitted directly opposite to maintain rotor balance.

The screen can be reversed to obtain maximum life. When the four corners of the hammers are worn, the two faces of the screen will also be worn out.

Any stones or other solid objects in the grain will be trapped in the hopper, and they should be removed after all the grain has been milled.

Using Hammer Mills

The mill must be running at full speed before grain is released from the hopper. The air used to convey the meal to the bin or mixer must be separated from the meal. This can be done in two ways. The first is the installation of a cyclone above the mixer. This is a conical-shaped container, at least 1 m in diameter, with the point at the bottom. When the meal and carrying air reach the cyclone, they whirl round inside the cone. The meal is thrown outwards and falls to the outlet at the bottom of the cyclone. The air is exhausted from the top of the cyclone and is piped from the building.

An alternative method of separating the carrying air from the meal is the use of filter bags on top of the storage bin or connected to a food mixer. The filter bags allow the air to escape and the meal is retained inside the filter.

Food Mixers

There are two types of mixer for the preparation of batch feeds for livestock. Vertical and horizontal batch mixers are permanently installed in a building. Mobile diet mixers mix bulk rations and discharge the mixed ration into livestock troughs for feeding.

Vertical Mixers

These have a large diameter vertical auger driven by an electric motor at 250–400 rpm. The auger runs inside a vertical tube in the centre of a cylindrical-shaped hopper with a conical base. The ground meal is either blown into the hopper from a hammer mill or tipped, together with the other ingredients, into the mixer hopper. The ingredients are mixed by the auger, which lifts them to the top of the tube and then allows them to fall back to the bottom of the mixer for recirculation. This procedure continues until mixing is complete, when the contents are discharged from a spout and either put into bags or conveyed to feed troughs.

Vertical mixers have a capacity of between 500 kg and 2,000 kg, although even larger models are made. Some mixers have an automatic mechanical or electronic weigh hopper that measures pre-set quantities of each ingredient and tips them into the mixer hopper. A typical 1 tonne mixer will mix a batch of feed in about 20 minutes.

Horizontal Mixers

Machines for farm use have a hopper capacity of between 500 kg–2,500 kg. The mixer hopper is filled by an auger or from bags, and the contents are mixed with a rotary agitator in the hopper bottom. A batch takes from 5–15 minutes to mix, after which the contents are emptied from a bagging-off chute into sacks or conveyed to feed troughs.

A similar type of mixer with an inclined hopper and a chain and slat conveyor mechanism (Figure 20.6) mixes the ingredients as they are carried to the top of the mixer. The ingredients then fall back to the bottom and are recirculated until the mix is complete. It takes between 5 and 10 minutes to mix a 1 tonne batch of feed. The completed mix is either bagged off from a spout or conveyed to the feed area.

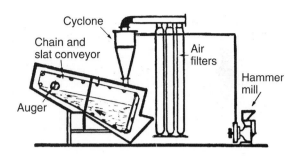

Figure 20.6   A batch mixer with a chain and slat conveyor. Meal is blown from the hammer mill to the mixer, where air filter bags and a cyclone separate the meal from the conveying air.

Mobile Diet Mixer–Feeders

These process and mix various feed ingredients to produce a total mixed ration, and are then towed alongside livestock feed troughs, where they discharge the completed mix. There are two main types: one mixes with horizontal paddles, and the other has one or two large cone-shaped vertical augers. Power take-off driven mixer-feeders can be used to chop and mix big bales of hay, straw or silage with molasses, roots and concentrated feeds. One or two heavy-duty augers with reversible chopper blades chop and mix the ingredients in the hopper, and when they are mixed, a chain and flight conveyor empties the hopper from a side, front or rear discharge spout.

Plate 20.10   Filling cattle troughs with a vertical auger diet feeder wagon. (Shelbourne Reynolds)

Plate 20.11   After loading a big bale into the hopper of this feeder/bedder, it is shredded and blown into feeding troughs. (Kuhn)

Mobile Straw Chopper–Spreaders

Mounted and trailed chopper blowers, either power take-off or hydraulic motor driven, are used to chop big round or big square bales and blow the chopped straw to a distance of up to 25 m into large livestock pens. The same machine can also be used to chop and blow big bale or bulk silage into a feeder wagon or feed troughs. A typical medium-sized trailed chopper-spreader, requiring a 60 kW (80 hp) tractor, will hold two round bales, a square bale or approximately 8 m^3 of clamp silage.

Another design of a mounted big round bale chopper blower can be used to shred and spread straw bales for bedding and chop hay or silage bales for feed. The bale is placed into a chain-driven rotary tub with a contra-rotating chopper rotor at the bottom of the tub, where it is chopped into short lengths and then discharged from a conveyor at one side of the tub. A tractor of about 52 kW (70 hp) is needed for the machine, which can also be used to shred small rectangular bales.

Plate 20.12   The power take-off driven drum on this round bale feeder/bedder is used to chop silage or straw bales and blow the chopped material into cattle pens. (Teagle)

DAIRY EQUIPMENT

Milking machines work in the same way as a calf sucks milk from a cow's udder. The calf withdraws the milk by frequently releasing the suction on the teat. This is done to allow blood to circulate in the udder and maintain the flow of milk.

Cows may be milked in a parlour or a cowshed, with a pipeline milking system or with robotic milking equipment. The cubicles in a parlour may be arranged abreast, tandem or herringbone, with staff working alongside the line of cows to attach the milking clusters. Alternatively, in a rotary parlour, the cows travel on a roundabout while they are milked and the milking staff stay in one place.

There are two systems of pipeline milking used for various designs of parlour. One system has a receiver jar that receives milk from the teat cups, and this is released at intervals to a milk pump that delivers it to a bulk tank in an adjoining milk room, where it is cooled. Another pipeline system has a large receiver vessel under vacuum created by a vacuum pump. The milk passes directly from the teat cups to this vessel and is then pumped to a bulk tank where it is cooled.

Robotic milking equipment carries out the complete milking procedure automatically. A sensor identifies the cow as it enters the milking cubicle from the information on an electronic chip on its collar. A robotic arm cleans the udder, puts on the teat cups, milks the cow and then removes the cluster.

The milking unit is operational for 24 hours every day, and the cow decides when she wishes to be milked, so the system is only suited to farms where the cows can walk directly from their grazing or feeding area to be milked. A

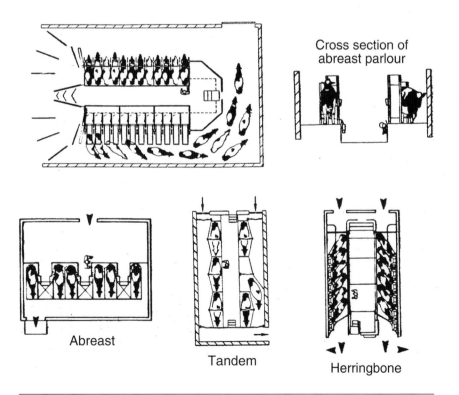

Figure 20.7   Examples of milking parlour layouts.

Plate 20.13   A robotic milking machine allows individual cows to choose when they want to be milked, and requires no human presence. (Lely)

Plate 20.14   A herringbone milking parlour with the cows standing at an angle in cubicles on each side of a central pit. The parlour can feature automatic feeding, cubicle gates and cluster removal. The operator stands in the well between the two rows of stalls to milk the herd. (Dairymaster)

Plate 20.15   Cows step onto this slowly rotating rotary milking parlour at a gated entrance, and step off again, fully milked, once it has completed a full revolution. (Fullwood)

computer linked to the robotic unit records the time of milking, the amount of milk given on each visit and the quantity of food consumed. A complete record of the yield throughout the lactation is also made.

Milking Equipment

The vacuum pump is at the heart of all milking installations. Driven by an electric motor, the pump, which is connected to the milking pipeline, creates a level of vacuum required to withdraw milk from the udder.

The vacuum regulator maintains the vacuum with a remote sensor in the milking pipeline, which is measured with a gauge on the pipeline and kept at a constant level. Normal atmospheric pressure is 1,000 millibars (1 bar), and any closed system with a pressure of less than 1 bar is under vacuum. Traditionally, vacuum was measured in inches of mercury (Hg), and the vacuum regulator was set to maintain it at a level of between 13 and 15 in. Hg in the pipeline. In metric terms, the vacuum level is maintained at between 400 and 500 millibars.

An interceptor vessel, situated in the pipeline near the vacuum pump, protects the latter by collecting any dirt in the pipeline which might otherwise damage the pump.

The sanitary trap is also included in the pipeline. Similar to the interceptor, it serves as a junction to link the vacuum pump to the vacuum and milk lines.

The pulsator on the vacuum line creates an intermittent vacuum between the liner and the teat cup shell. Various types of electronic and pneumatic pulsators are used to set the pulsation rate. This is the number of milking and rest phases per minute. It will vary with different types of milked animals and is designed to reproduce the natural sucking rate of a calf or, in the case of milking goats, the kid.

The pulsator also controls the pulsation ratio, which is the ratio of the length of the milking period to the length of the rest period. The pulsation rate is normally between 50 and 60 cycles per minute, and the pulsation ratio may be 50:50, 60:40 or 70:30. The first figure refers to the milk withdrawal period and the

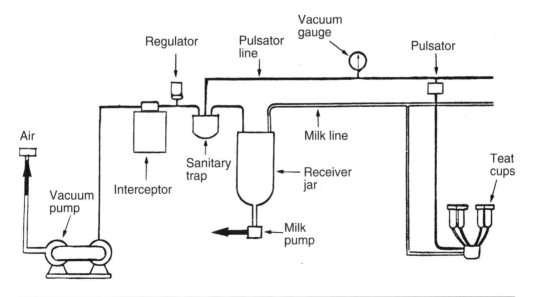

Figure 20.8   A pipeline milking system.

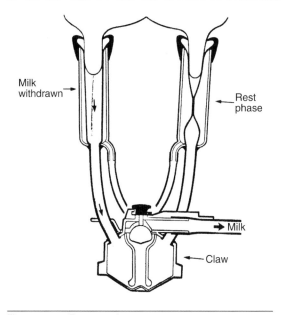

Figure 20.9   Teat cup cluster.

second to the rest period. Some pulsators are set to give a rest period to all teat cups at the same time, while others have two at rest and two in the milk phase.

The teat cup cluster consists of a central claw with tubes connecting it to the teat cups. Teat cups have either a metal or plastic shell with a rubber liner. One tube on each teat cup is connected to the milk line and is under constant vacuum. The second tube connects the vacuum line to the space between the liner and the teat cup shell and has intermittent vacuum.

Milk is withdrawn from the teat when there is vacuum on both sides of the teat cup liner. The pulsator releases the vacuum between the liner and the shell at intervals, and this causes the liner to collapse inwards and squeeze the teat. This is the rest phase, and the milk flow is cut off.

Automatic teat cup removal is a feature of many milking parlours. A sensor measures milk flow, and when it stops a valve cuts off the vacuum and the cluster is removed automatically. The cluster is suspended from a cord linked to the vacuum line, and when the vacuum is cut off, the cluster is automatically lifted away from the cow.

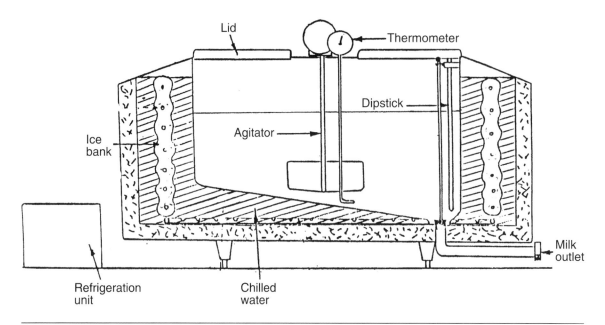

Figure 20.10   Refrigerated bulk milk tank.

Milk Recording

The amount of milk given by individual cows during a lactation can be recorded automatically by noting markings on individual glass receiver jars before it is released to the milk pump and the delivery line to the milk tank. Jars have become far less common in modern milking parlours, which are now usually fitted with automatic flow rate recorders in the pipeline from the teat cup cluster. This is linked to an electronic recording system, which compiles yield information for each cow.

Cooling

In an older type parlour with receiver jars, the milk is pumped at intervals, under vacuum, to the milk tank, while in a more modern parlour with a direct milk line to the tank, the flow is constant. Before it reaches the tank, the milk is passed through a filter. Then, having reached atmospheric pressure, it is cooled from about 35° C to between 4 and 6° C and stored in a refrigerated bulk tank. A mechanical paddle in the tank intermittently stirs the milk to speed up the cooling process and evenly distribute the butterfat, or cream, which would otherwise all gather on the surface. It is particularly important that the milk is stirred for a short spell just before it is collected.

Cleaning

Milk pipelines and utensils must be washed after use to prevent contamination and maintain good dairy hygiene. Circulation cleaning and sterilisation is an automatic feature of all modern pipeline milking installations. Many dairies have a computerised monitoring system that not only checks the performance of each cow and the condition of the milk, but also checks all aspects of cleaning and sterilising the milk lines and tank.

ELECTRIC FENCERS

Mains and battery operated electric fencers can be used to fence off permanent paddocks or divide pastures for temporary grazing. They also have uses for other livestock, such as temporary outdoor pig enclosures that may be moved every few years.

Fencer units may be powered by a 9 or 12 volt battery, mains electricity or solar power. The fencer unit transforms electricity supplied by a battery or mains supply to a much higher voltage – between 5,000 and 10,000 volts, depending on the model of fencer. An electric fencer works in a similar way to an engine coil ignition system. A transistorised circuit breaker interrupts the flow of current in the primary low voltage circuit, resulting in a high voltage current being induced in a coil in the secondary circuit. There are between 40 and 60 pulses of high voltage current produced per minute. These high voltage pulses at a very low amperage pass through the fence wire, and any animal touching the wire will earth out the current and receive a short sharp shock.

The fence wire, or high visibility wire-impregnated tape for horse paddocks, is supported by plastic or porcelain insulators attached to plastic, wooden or metal posts. A special type of polythene mesh net with wire woven into it or several strands of wire may be used for small livestock. A host of available fencing accessories include gate handles, strainer posts and neon testers to check the fence is working properly.

Tall vegetation must not be allowed to touch the fence as it will partially earth out the current and reduce the effectiveness of the fence.

CHAIN SAWS

Chain saws are the most dangerous hand tools on the farm and should be handled with great care. They can only be used right-handed and they should not be used without first receiving adequate training and passing the relevant Certificates of Competence (see page 260). The Health and Safety at Work Regulations also require operators to wear full protective clothing when using a chain saw.

Plate 20.16   The chain safety brake on this chain saw, immediately in front of the main handle, stops the chain immediately if the bar kicks back towards the operator.

Most chain saws have a two-stroke petrol engine with a diaphragm carburettor that allows the saw to work in any position. Some small chain saws have an electric motor. When it is at full working speed, the engine runs at 9,000–15,000 rpm depending on model, with some specialist saws running even faster. The cutting chain runs at speeds of 95 kph (60 mph) or more.

Chain saw guide bars are made in various lengths, ranging from 300 mm on electric saws up to 600 mm on a large petrol-engined model. Longer guide bars are available, but they are mainly used by tree surgeons. A typical chain saw for general farm work has a guide bar of between 380 and 540 mm in length and an engine in the 2–3.5 hp (2.8–4.8 hp) bracket.

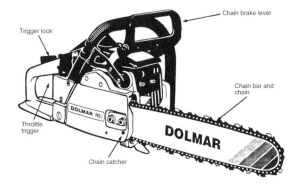

Figure 20.11   Chain saw controls.

Controls

Recoil starter The starter cord should be pulled slowly until compression resistance is felt. Then apply a sharp pull. Do not pull the starter cord to its full extent, or suddenly release it; the cord should be allowed to rewind slowly.

Throttle trigger This controls the speed of the cutting chain. It cannot be used until the trigger lock is depressed. When the engine reaches a pre-set speed, the centrifugal clutch engages the drive to the chain. At engine tick over, the cutting chain will be stationary.

On/off switch This isolates the ignition system so that the engine will neither start or run.

Chain brake lever This is a safety device which will stop the cutting chain instantly when it is pushed forward with the left hand. The chain brake should be tested before using the saw. The chain brake also acts as a guard to protect the operator's hand if it accidentally slips off the handle.

Chain saws have an automatic inertia-type chain brake. This engages automatically when kickback occurs without the need to operate the chain brake lever. Kickback is the uncontrolled upward movement of the outer end of the guide bar if it comes into sudden contact with an obstruction or if the bar becomes trapped in a saw cut. This results in upward movement of the saw and brings the chain brake lever up against the back of the operator's hand and stops the chain.

Safety devices
In addition to the chain brake, other safety devices to protect the operator include:

- *Rear handle guard* Protects the right hand if the chain breaks.
- *Chain catcher* A pin which catches the chain if it breaks. It works in conjunction with the rear handle guard.
- *Anti-vibration system* This dampens saw vibration, which reduces operator fatigue and improves cutting efficiency.
- *Safety lock-out switch* Isolates the throttle trigger so that the cutting chain cannot be accidentally engaged.

Chain saw users are required by law to wear a safety helmet, ear defenders, eye protection, protective gloves, trousers made from a special material which will clog the cutting chain and chain saw safety boots. Clothing should be close fitting.

Maintenance
The most important factor in chain saw operation is to keep the cutting chain sharp. It is cutting correctly when the saw throws out chips rather than dust. The latter indicates that the chain requires sharpening.

Frequent sharpening is required, but only two or three strokes with a chain saw file are needed to bring a well-maintained cutting chain back into full efficiency. Chain tension is important. A loose chain might come off the guide bar, and a tight one will cause excessive wear of the guide bar and drive sprocket. An adjuster screw is provided to set the chain tension. It is good practice to turn the guide bar at regular intervals to even out the rate of wear.

Clear wood chips away from the chain brake mechanism and make sure it works correctly.

An automatic chain lubrication system is standard on all but the most inexpensive electric chain saws that have a hand-operated oiler to lubricate the cutting chain. For both lubricating systems, the oil reservoir must be topped up with chain saw oil to ensure that the chain

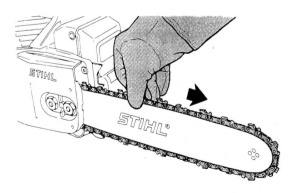

Figure 20.12   Checking the tension of the cutting chain.

does not run dry on the guide bar. Always use chain saw oil that has special additives to help keep the oil on the cutting chain. To check for efficient lubrication, run the saw with the end about 150 mm above a piece of wood. An increasing film of oil on the wood indicates correct lubrication of the cutting chain.

The engine will need regular maintenance, with particular attention paid to the air cleaner and sparking plug. A blocked air filter will cause excessive exhaust smoke and choke the engine. A dirty sparking plug is a frequent cause of poor starting. It is a good idea to have a spare plug carefully stored in the tool kit.

Keep the engine-cooling fins clean. Fins clogged with wood chips will cause the engine to overheat.

Using Chain Saws

Health and Safety Regulations state that any person using a chain saw must be over 18 years of age, have received training and hold appropriate Certificates of Competence for the work undertaken.

Dealing with hung and wind-blown trees is a dangerous operation and should be left to an experienced and suitably qualified person.

Before using a chain saw, make sure the safety devices are working and the cutting chain is correctly tensioned. Start the saw on firm ground with the right foot on the rear handle and the left hand on the front handle. Engage the throttle, set the choke on a cold engine, switch on the ignition and pull the starter cord. When it starts, run the saw and check the operation of the chain brake. Make sure the area is clear of obstructions and all bystanders before using the saw. Set the saw at full throttle before starting to cut.

HEDGE CUTTERS

There are two types of hedge cutter for farm hedges. Flail hedge cutters with a high-speed swinging flail rotor are able to deal with any thickness of growth up to small trees, and are

in almost universal use. The rotor can be positioned to cut the tops and sides of hedges and the sides of banks. Reciprocating knife hedge cutters, now rarely used, have a heavy-duty cutter bar that can be angled with a hydraulic ram. The thickness of material the bar can cut is limited.

Plate 20.17   Suitable for cutting the top and sides of a hedge, this 5.2 m reach flail hedge cutter has a 1.2 m wide cutting flail rotor. (Twose)

Plate 20.18   Hedge cutters can also be used to flail verge growth on road sides. (Bomford)

Flail Hedge Cutters

A hydraulic motor drives the cutting rotor on a flail hedge cutter and hydraulic rams are used to angle the cutting head. A typical flail hedge cutter has a rotor speed of 3,000 rpm for hedge cutting and 2,000 rpm for grass banks. A rear roller is sometimes used when cutting at ground level and special grass cutting flails should be used for verge trimming, etc. Many flail hedge cutters have an independent hydraulic system with an oil reservoir on the machine and a pump driven by the tractor power take-off. Others use the tractor's hydraulic system. The large tank on the right-hand side on the hedge cutter illustrated in Plate 20.17 is the oil reservoir for the machine's independent hydraulic system.

Most hedge cutters have a rotor with a cutting width of 1–1.5 m. The reach – that is, the maximum distance from the tractor to the flails – depends on the length of the flail rotor arm. A typical general-purpose flail hedger, with a two-part arm, has a reach of about 5 m.

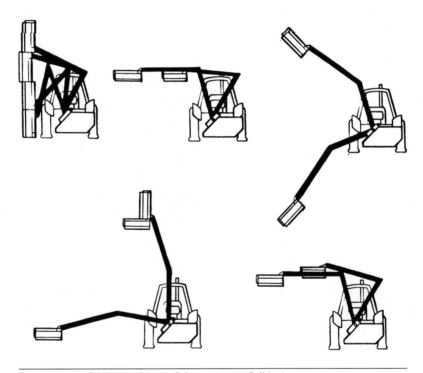

Figure 20.13   Cutting positions of the rotor on a flail hedge cutter.

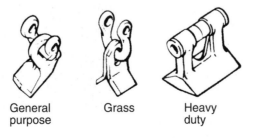

General   Grass   Heavy
purpose   duty

Figure 20.14   Types of hedge cutter flail.

Some hedge cutters with a three-section flail arm have a reach in excess of 7 m. Flail hedge cutters designed for use in the UK cut on the left of the tractor as it travels on the left in the direction of the traffic when cutting from the road. Some flail hedge cutters can be slewed with a hydraulic ram to cut on either side of the tractor. The position of the rotor head (Figure 20.13) is adjusted on most machines by altering the angle of the two or three sections of the flail arm, while other machines have a ram-operated telescopic outer section. A breakaway device is built into the flail arm to protect it from damage if it hits an obstruction.

DITCHERS

Some farms use an industrial tractor digger, more commonly seen on building sites, with a heavy-duty front loader bucket and a backhoe at the rear for digging trenches. Alternatively, farmers may choose to hire a 360-degree excavator, or employ a contractor to do the ditching or digging work on the farm. There are also various types of three-point linkage mounted back hoe / ditch digger operated by the tractor hydraulic system, although these are now less common. Others have an independent hydraulic system with a power take-off driven pump and an oil reservoir attached to the digger frame. Digging buckets on all of these machines range from very narrow ones used to dig trenches for water pipes up to ditch cleaning buckets with a width of 1.5 m or more.

When in work, tractor-mounted machines are supported by hydraulically raised and lowered legs. A set of control levers is used to move the ditcher arm up and down and from side to side through a working arc in excess

Plate 20.19 Cleaning out a ditch with a bucket mounted on a backhoe. (Caterpillar)

of 180 degrees. The working position of the two-part arm, above and below ground level, is controlled with two rams, while a third ram operates the digging bucket. A typical farm ditcher can excavate a trench up to 2.5 m deep and has a reach of 4.25 m from the support legs to the tip of the bucket.

IRRIGATION

Yields from some arable crops, particularly root crops and field scale vegetables, will be reduced if they are short of water, especially when grown on light soils. Fruit and grass grown in drier areas will also benefit from irrigation. The irrigating water is taken from man-made reservoirs or boreholes or from rivers, and a large pump distributes it to the fields through underground mains or a network of semi-permanent surface mains.

Travelling rain guns linked to large hose reels and rotating sprinklers attached at intervals to a portable pipeline are the two main systems used to irrigate farm crops.

Mobile Irrigators

Self-propelled mobile reel irrigators have the advantage of not requiring a portable aluminium

Plate 20.20   The rain gun trolley is pulled back across the field by the hosepipe as it is gradually rewound on the mobile hose reel. (Bauer)

pipeline needed for sprinkler systems. Mobile reel irrigators have a large trolley-mounted hose reel with a long length of hosepipe and a smaller trolley with either a rain gun or a rotary sprinkler. With the irrigator hose reel connected to a water main at one side of the field, the rain gun trolley is towed to the opposite headland and the hose unwinds from the reel.

When irrigation commences, a turbine driven by the irrigating water turns the hose reel very slowly and gradually rewinds the hose. Water is pumped from the rain gun or sprinkler as the trolley is pulled back across the field. There are various models of mobile irrigator, some with up to 400 m of hose, and most irrigators have a variable hose rewind speed to vary the amount of water applied to the crop.

Sprinkler Irrigation

Portable aluminium mains with water- or impact-driven sprinklers on short riser pipes from the aluminium main are preferred for irrigating smaller areas of crop. Portable aluminium spraylines, usually 100 mm in diameter and either 6 or 9 m in length, are connected together with quick release couplings that allow speedy assembly and dismantling. Impact sprinkler nozzles are used for most field scale pipeline irrigation systems. The normal working height of a sprinkler unit is between 600 and 1,000 mm

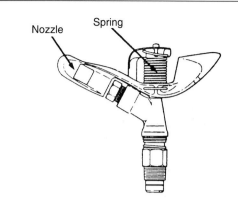

Figure 20.15   A full circle impact sprinkler unit.

above the ground level, depending on nozzle output; and it is usual for the sprinklers to be spaced at 9 or 12 m intervals along the spraylines.

Impact sprinkler nozzles are used for most field scale pipeline irrigation systems. The nozzle has a spring-loaded arm (Figure 20.15), which is pushed sideways by the pressure of the water as it leaves the nozzle. Spring pressure returns the arm to its rest position, and in so doing it slightly rotates the nozzle unit. This cycle is repeated continuously, causing the nozzle to rotate many times through 360 degrees to irrigate a complete circle of ground.

CHAPTER 21

Farm Power

Various sources of power are available on most farms. Internal combustion engines provide the power for working the land, and electricity, oil and gas are the major sources of energy for livestock buildings, drying grain and the farm workshop.

Plate 21.1   Both shown here, internal combustion engines and electricity are two important sources of power on the farm. A diesel engine provides the power for the tractor, while electricity is supplied to farms from the high voltage power line on the far side of the field. (Lemken)

INTERNAL COMBUSTION ENGINES

All modern farm tractors, self-propelled machines and standby generators have diesel engines, while petrol engines are used to drive such equipment as chain saws, brush cutters, small standby generators and other farmyard equipment.

Engine power is measured in horsepower (hp) or in kilowatts (kW). Tractor sales literature may quote engine power in horsepower or kilowatts and some give both figures. One horsepower is equivalent to 0.75 kW. This means that a 100 hp engine is also rated at 75 kW.

Brake Horsepower (BHP) or Kilowatts

Often shortened simply to horsepower or hp, this is the power available at the engine flywheel. It is measured with the engine on a test bed with a dynamometer. This device applies a type of brake to the engine flywheel and measurements taken during the test are converted to brake horsepower. Sales leaflets always quote BHP at rated speed, the engine speed chosen by the manufacturer, and this is around 2,000 rpm. The rated speed is lower than the maximum engine speed.

Drawbar Horsepower

DBHP is the power available at the drawbar. It will always be less than brake horsepower because some of the engine power is used to drive the transmission, hydraulic system and other components.

Power Take-Off Horsepower

PTO HP will be less than brake horsepower but greater than the drawbar horsepower. It is measured by connecting a dynamometer to the power take-off shaft. Large power take-off driven machines have a high power requirement. As a general guide, a tractor will produce about 80 per cent of its rated engine power at the power take-off shaft.

Many tractor sales leaflets list all three tractor power ratings. The figures may be given in horsepower or kilowatts, and in some cases both units are used. The following approximate power output figures at rated speed are typical for a medium-sized farm tractor:

- Engine power   100 kW or 135 BHP
- Power at the power   90 kW or 120 PTO
  take-off shaft   HP
- Power at the tractor   70 kW or 94 DBHP
  drawbar

These figures show that a 100 kW (135 BHP) tractor has only 70 kW (94 hp) at the drawbar to pull an implement. This amount of power at the drawbar will be reduced still more due to power lost through wheelslip, rolling resistance and flexing of the tyre walls. Rolling resistance is experienced when pushing a wheelbarrow on soft ground, which makes it hard work because the wheel sinks into the soil. Rolling resistance on tractors is increased on soft or previously cultivated ground, and tractor front wheels have to overcome the same resistance.

Tractor sales leaflets may give engine power in horsepower or kilowatts qualified with an ISO, BS AU, DIN or ECE rating. The letters ISO 14 396 and BS AU 141a indicate that this is the maximum gross power available at the flywheel calculated on a test bed with none of the engine power lost in driving an alternator, water pump, cooling fan, etc.

DIN 70 020 and ECE R24 are the net maximum power outputs available at the flywheel. This will always be lower than gross engine power because the engine is driving the alternator, water pump and other auxiliaries while under test.

Some tractor sales literature gives engine power in cv or ps. The French cv is an abbreviation for *chevaux-vapeur* and ps is short for the German *pferdestärken*. Both are metric horsepower ratings, with 1.014 cv or 1.014 ps equal to 1 hp.

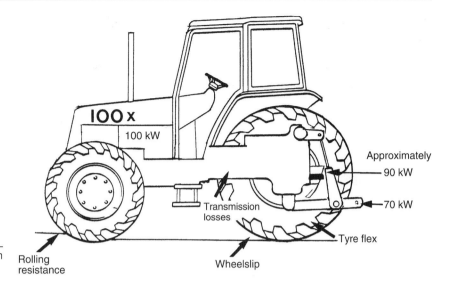

Figure 21.1 Power losses in a tractor.

ELECTRICITY ON THE FARM

Internal combustion engines can be permanently installed inside a building to drive fixed equipment, but hot surfaces must be guarded and the exhaust fumes must be piped to the outside of the building.

None of these requirements apply to an electric motor. In fact, the opposite applies because electric motors must not be exposed to damp conditions unless they are specially designed for use in these conditions.

Electricity is a major source of farm power. It is generated by a network of power stations that feed electricity into the mains supply lines (the National Grid) at a pressure of 275,000 or 400,000 volts. This is reduced to 132,000 volts and again to 33,000 volts at transformer substations. A village may have a supply of 11,000 volts from a transformer on a 33,000 volt line. Pole transformers reduce this to 240 volt single-phase or 415 volt three-phase supply for farm and domestic use.

The mains electricity supply, usually 240 volt single-phase transmitted by two wires, is adequate for some farms. Three-phase supply is a combination of three separate single-phase supplies arranged to give a pressure of 415 volts, which is an advantage when using grain handling, milling and food mixing equipment, which have a high power demand. A three-phase supply is very expensive to install and it is not available in some rural areas.

Mains electricity is alternating current (AC), which cannot be stored in a battery. In contrast, direct current (DC) generated by a tractor alternator is stored in a battery for use when required.

Electrical power is measured in watts, and there are 1,000 watts in one kilowatt. The rate of flow of an electrical current is a circuit measured in amps. Electrical circuits and appliances have fuses to protect them from damage through overloading. The size of fuse in amps governs the amount of electricity flowing in the circuit. The current flow in a circuit can be calculated with this formula:

$$\text{Amps} = \frac{\text{watts}}{\text{volts}}$$

For example, an electric heater connected to a 240 volt supply has a power rating of

1,000 watts. The current flow is calculated in this way:

$$\text{Amps} = \frac{\text{Watts}}{\text{Volts}} = \frac{1{,}000}{240} = 4.16 \text{ amps}$$

When the number of amps and volts is known, the formula can be turned round to calculate the watts in a circuit:

Watts = amps × volts

For example, to find how many watts can be carried in a 240 volt circuit with a 5 amp fuse:

Watts = amps × volts = 5 × 240 = 1,200 watts

In practice, a 5 amp fuse would be used for a 1,000 watt electrical appliance.

Most portable mains electric equipment has a fused plug. It is important to have the correct fuse for the equipment, and the above formula can be used to find the correct fuse rating in amps provided that the voltage and watts are known.

Fuse rating	Maximum loading (watts)	Typical use
2 amps	500	Inspection lamp, soldering iron
5 amps	1,200	Small bench grinder, Hand electric drill
10 amps	2,400	Heavy-duty power tools, angle grinder
13 amps	3,000	Small portable electric welder

These fuse ratings apply to single-phase 240 volt equipment. Three-phase power tools and equipment are wired into an isolator switch and have a fuse in the main supply fuse board.

Replacing a Fuse

When you need to replace a fuse:

- Always disconnect the power supply first.
- Replace the fuse with one of the same rating.
- Consult an electrician if the fuse blows again as soon as the power is turned on.

Residual Current Devices

Instead of replaceable fuses, many electrical installations have residual current devices (RCDs) on the 240 volt distribution board. When an overload or dangerous situation occurs, the RCD cuts off the supply in milliseconds, giving far greater protection to the equipment and the user. When an RCD cuts off the supply of electricity, disconnect the appliance before resetting the RCD. If the RCD fails to reset or trips out again when the appliance is used, it should be checked by a qualified electrician. RCDs can be checked with a test button on the distribution board. Read and follow the instructions supplied with the RCD.

Care of Electric Motors

Totally enclosed electric motors are used for equipment that has to work in dusty conditions. Some electric motors in the farm workshop and food preparation areas do not have full protection from dust. In such cases, give what protection you can to stop dust collecting inside the motor, as this will restrict air circulation and in time the motor may overheat. Attention to the following maintenance points will help to prolong the working life of an electric motor:

- Wipe off any oil which may appear or be spilled on the motor. It will attack the insulation and attract dust.
- Keep the motor dry.
- Avoid overloading the motor, as this will cause it to overheat.
- Make sure the motor is secure in its mounting bracket. Vibration increases wear.
- Check that the belt pulleys on the motor and the machine are in alignment. Misalignment will result in worn bearings and a worn drive belt.

Wiring a Plug

Follow the colour code. Always connect the three wires in an electric cable in this way:

- *Green and yellow* to the largest plug pin. This is the *earth* connection.
- *Brown* to the *live* terminal. This terminal is connected to the fuse holder in a fused plug.
- *Blue* to the third *neutral* terminal.

To wire a three-pin plug:

1. Strip about 50 mm of the outer covering from the cable.
2. Hold the unstripped end of the cable against the cable grip and run the three coloured wires to their correct terminals. Cut each wire to leave about 10 mm of cable beyond the terminals for making the connections.
3. Remove about 10 mm of covering from each of the coloured wires.
4. Bend the wires clockwise round the terminals so they stay in position when tightening the terminal screws. Modern terminals have a hole in the top of the pin for the wire with a screw to secure it. Cut off any ends of bare wire not needed to hold it in the terminal.
5. Fit the correct fuse and secure the cable in the plug with the cable clip. Check your work before replacing the plug top.

Some power tools have only two wires: one blue and the other brown. There is no earth wire because the appliance is double insulated.

THE SAFE USE OF ELECTRICITY

Electricity can kill. The following points must be remembered when using or working near electrical equipment.

- Always use the correct fuse for the appliance.
- Protect light bulbs in farm buildings. Keep them clean and away from combustible material.
- Keep extension leads, plugs and sockets in good condition and locate them in a position where they will not be damaged.

- Never use unsafe electrical equipment. Report any faults to someone in authority as soon as possible.
- Do not attempt temporary repairs to electrical equipment. Call in a qualified electrician.
- Never work with electrical equipment with wet hands.
- Water is a conductor of electricity. Never tackle a fire caused by an electrical fault with a water-based extinguisher or any other type which uses any liquid. A dry powder extinguisher should be used to tackle an electrical fire.
- Overhead cables in the farmyard or across fields must be treated with respect. Accidental contact with overhead lines with metal ladders, high tipping trailers or irrigation pipes can have fatal consequences. When working near electricity posts with a tractor, beware of catching the post or a stray wire, as this might bring down both post and wires. Be aware that if this happens, the cable may still be live.

GAS AND OIL HEATING

Many farmers use electricity for heating. It is clean and trouble-free until there is a power cut. It can also be one of the more expensive forms of heating.

Oil is an alternative source of fuel, especially for providing the heat for the farm grain dryer. Provided that the burners are serviced at regular intervals and fire precautions are observed, oil, used as a heating fuel, presents few problems.

Some farmers favour gas, especially those who keep livestock that requires heated housing, such as young pigs and poultry. It can be collected in cylinders from a local dealer or delivered in bulk to the farm, where it is kept in permanent storage tanks.

Butane and propane, both heavier than air, are in common use. Permanent propane storage tanks must be sited outside buildings, but portable appliances including heaters can have the gas cylinder inside the building or workshop. Take care to keep the threads clean when connecting a gas cylinder to an appliance. Any leak will smell, and if you suspect a leak, turn off the supply at once. The leak can usually be found by applying a small amount of soapy water to all the connections; tell-tale air bubbles will come from a leaking connection.

TRANSMISSION OF POWER

The power developed by an internal combustion engine or an electric motor can be transmitted in various ways, including the following.

Vee-Belts

Vee-belts are widely used to transmit drive on various farm machines and equipment in the grain store and farm workshop. Vee-belts transmit power with the sides of the belt, which grip against the sides of the vee-belt pulley. The belt must not be allowed to run on the bottom of the pulley. Crossing the belt in the shape of a figure eight reverses the drive. Vee-belts must be correctly tensioned. This is done either with a jockey pulley or by adjusting the position of one of the pulley shafts to tighten the belt. A jockey pulley is a small pulley on an adjustable arm that can be moved to take up any slack in the belt. Vee-belt drives on implements and fixed equipment must be fully guarded at both run-on points and along the entire length of the belt. There must be an isolator switch or engine stop control at the working position of fixed equipment so that the operator can stop the drive to the machine in an emergency.

Vee-belts are sometimes in sets of three or four running side by side in multiple vee-belt pulleys. A tractor-mounted rotary grass mower, for example, has a multiple vee-belt drive to the cutting discs. Multiple drive vee-belts must be replaced in complete sets when they are worn.

Vee-belts, which will slip if they are overloaded, reduce the risk of serious mechanical damage to machinery and equipment.

Chain Drives

A chain drive provides a positive drive with no possibility of slip. Chain drives on farm machinery often have a slip clutch to protect the machine and the drive from damage through overload.

Roller chains are made in many sizes and also as a single, double row (duplex) and three-row (triplex) chain. Most chains are used for high-speed power transmission, but low-speed chains are also made. Roller chains need frequent lubrication to ensure a long working life. The wheels which carry the chains are called sprockets. The chain and the sprockets will wear at approximately the same rate. A worn chain has a lot of movement between the links, and worn sprockets will have hook-shaped teeth.

Chain drives must be correctly tensioned. This is done in the same way as tensioning a vee-belt drive with a jockey or idler sprocket, or by adjusting the position of one of the sprocket shafts. When all adjustment has been used and the chain is still slack, a link must be removed from the chain.

Shafts

A shaft is a simple method of transmitting drive over a distance. When the ends of the shaft are not contained with a rigid framework, allowance must be made for any misalignment of the shaft. This is normally achieved with a universal joint, which allows movement in all directions. A tractor power take-off shaft has a universal joint at both ends which allows the input shaft on the implement to move in relation to the tractor power shaft.

Gears

Spur gears, with straight or helical teeth, are used to transfer drive from one shaft to another running parallel to it. Gears with helical teeth (Figure 21.2) give a stronger and quieter drive

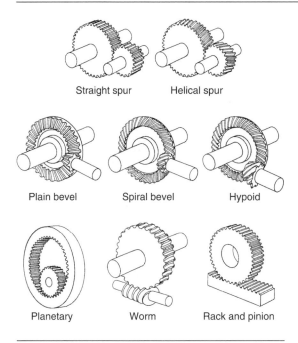

Figure 21.2   Types of gear.

than those with straight teeth. Gears can be used to increase or reduce the speed of the driven shaft, and a gear turning clockwise on the driving shaft will turn a gear on a parallel shaft in the opposite direction.

Bevel gears, also with straight or helical teeth, are used to transmit drive through an angle from one shaft to another. The crown wheel and pinion in a tractor transmission system is an example of bevel gears transmitting drive through 90 degrees. A worm and wheel drive is another way of transmitting drive through a right angle.

Gear and Pulley Speeds

The speed of a driven gear can be calculated if the speed of the driving shaft and the number of teeth on both gears is known.

Use this formula:

$$\frac{\text{Teeth on gear A}}{\text{Teeth on gear B}} = \frac{\text{rpm of gear B}}{\text{rpm of gear A}}$$

The same formula can be used for belt pulleys by substituting the number of gear teeth with the working diameter of the pulleys.

Example: Find the speed of gear A with 15 teeth driven by gear B with 6 teeth running at 80 rpm.

$$\frac{\text{Teeth on gear A}}{\text{Teeth on gear B}} = \frac{\text{rpm of gear B}}{\text{rpm of gear A}}$$

$$\frac{15 \text{ teeth}}{6 \text{ teeth}} = \frac{\text{rpm B}}{80 \text{ rpm}} = \frac{15 \times 80}{6} = \frac{1200}{6} = 200 \text{ rpm}$$

Safety Clutches

Chain drives and some belt drives are protected by a safety clutch that will slip if the drive is overloaded.

One type of slip clutch has two serrated metal faces held together by spring pressure. Figure 21.3 illustrates a vee-belt pulley with this type of slip clutch. When overloading occurs, the pulley continues to turn, but the shaft does not, and the movement of the driven face against the stationary face makes a loud clattering noise. This type of safety clutch is also known as a clatter clutch.

Another type of slip clutch has one or two discs lined with a friction material and held

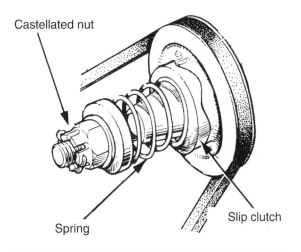

Figure 21.3   Serrated face slip clutch.

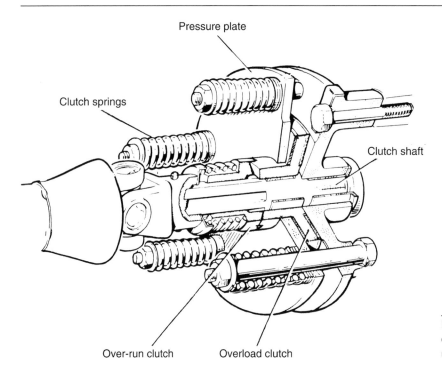

Pressure plate

Clutch springs

Clutch shaft

Over-run clutch   Overload clutch

Figure 21.4   A friction disc slip clutch and over-run on a pick-up baler flywheel.

by spring pressure between smooth pressure plates attached to the power shaft. The friction disc or discs are fixed to the driven shaft, and when overloading occurs, the power shaft still turns but the friction disc is stationary. There is no noise from the clutch, but with continuous slipping it will overheat.

Some power take-off driven machines, including pick-up balers, have an over-run clutch which acts in the same way as the free-wheel on a bicycle. The clutch is spring-loaded and has one plate with a square jaw and one with angled jaws. The square jaws drive the machine when the power shaft is engaged. When the drive is disconnected, the over-run clutch comes into action, and the spring allows the plate on the driven shaft with its angled jaws to continue turning and slow down at its own pace. An over-run clutch makes a clicking sound as the plates rub against each other.

Hydraulics

Remote hydraulic rams, connected to the trac-tor hydraulic system, are used to raise and lower loads, reverse plough bodies and make

all manner of working adjustments on farm machinery. A single-acting ram used, for exam-ple, to raise a tractor front-end loader, only applies force in one direction on a simple older loader. It relies on the weight of the loader to lower it again. A double-acting ram with two pipes connecting it to the tractor hydrau-lic system applies force in two directions. A double-acting ram is used, for example, when reversing the bodies on a reversible plough.

Hydraulic motors, connected by plug-in hoses to a tractor hydraulic system, provide constant hydraulic power without the need for any mechanical connection. Hydraulic motors are an extremely flexible method of transmit-ting power and can be used in any position on trailed and self-propelled farm machinery. Examples of their use include hedge cutters, fertiliser broadcasters, grain drills and com-bine harvesters. With prolonged use, the oil used by a hydraulic motor will become very hot, but engineers have designed a number of ways to keep the oil below its critical tem-perature such as a separate radiator to cool the hydraulic oil.

CHAPTER 22

The Farm Workshop

Every farm has a workshop. It may be a building set aside for the purpose or a bench at the back of the tractor shed. Even the smallest workshop needs a set of hand tools, an electric drill and a grinder to help keep a keen edge on a variety of cutting tools. Many farms have a well-equipped workshop with enough hand and power tools and a welding set to carry out a wide range of construction and repair work. The following pages describe some of the more basic hand and power tools found in many farm workshops.

WORKSHOP HAND TOOLS

Spanners

There are several types of spanner in common use, but spanner sizes can be a mystery to the occasional mechanic, who may prefer to take the easy way out with an adjustable spanner.

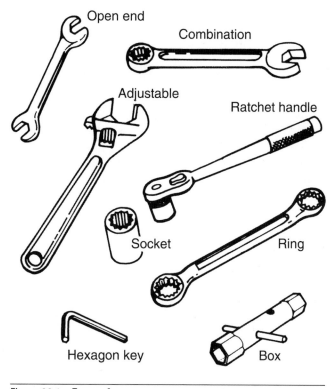

Figure 22.1   Types of spanner.

Open ended spanners, or flat spanners, are part of every tool kit. However, they may not be suitable for undoing nuts in confined spaces. Open ended spanners may also slip off the nut or bolt head if used carelessly.

Ring spanners fit around the nut or bolt head and are less likely to slip off. They are ideal for turning a nut in a confined space, as only one-twelfth of a turn is necessary before the spanner can be repositioned on the nut. Ring spanners may have flat or cranked ends, which makes them suitable for use on nuts in a small recess. Both ring and flat spanners have a different size at each end.

Combination spanners have an open jaw at one end and a ring at the other; the open jaw and the ring are both the same size on these spanners.

Socket spanners are more expensive but have many advantages, especially their speed and the ability to reach nuts or bolt heads in recessed holes. They can be used with a range of handles, including a speed brace, ratchet, tee-handle, a breaker and some extension bars, which can be attached to the socket and used to reach nuts inside a component or other awkward places. A universal coupling on the handle allows it to be used at an angle to the socket. The standard range of sockets has a ½-in. square drive, but larger and smaller square drive sockets are also made. They range from miniature sockets with a ¼-in. square drive to heavy-duty sockets with a ¾-in. or 1-in. square drive. (Tool catalogues use imperial measurements for square drive socket sizes.)

Bi-hexagon sockets, like ring spanners, have twelve faces and can be used to make one-twelfth of a turn when working in a confined space. Although less common, single hexagon sockets with six faces will give a better grip on very tight nuts and bolt heads.

Box spanners, often called plug spanners, are a cheap but effective form of socket spanner. Unlike bi-hexagon sockets, they have six faces and are turned with a steel rod or tommy bar.

An adjustable spanner may be a mechanic's friend, but when the jaws become stretched and worn it should put in the scrap bin. Worn jaws will wear the corners from or slip off the nut, sometimes with painful consequences. Adjustable spanners are made with different length handles and maximum jaw opening widths.

Torque spanners are used to tighten nuts or studs to a precise setting or torque. Used with a socket, a torque spanner is a mechanic's tool used, for example, to tighten cylinder head studs after an engine overhaul.

Hexagon keys, also known as Allen® keys, are 'L' or 'T' shaped. They are used for screws and bolts with a hexagonal recess in the head.

Spanner sizes

The size of a spanner relates to the distance across two opposite faces of a nut or the type of thread used for the nut or bolt. Coarse threads are stronger than fine threads, but fine threads are less likely to vibrate loose in service. The main spanner sizes are:

Metric Spanners Metric threads have been used for most British-made tractors and machinery for the last thirty years or so. Spanners for metric nuts and bolts are marked with the width across the opposite faces or flats of a nut, which is also the measurement between the spanner jaws. A metric spanner marked '22' fits a nut or bolt head measuring 22 mm across the flats (see Figure 22.2).

Although metric threads are in universal use, some elderly tractors and farm machines will have nuts and bolts with British or American thread rates.

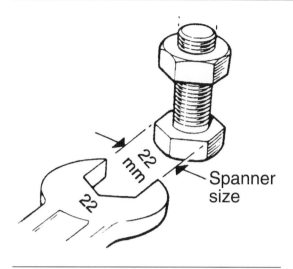

Figure 22.2   Spanner sizes.

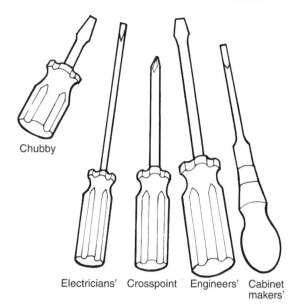

Figure 22.3   Types of screwdriver.

A/F (Across Flats) spanners These are used for nuts and bolts with an American National Coarse or Fine (ANC or ANF) thread. The spanner is usually marked A/F and the size refers to the width across the flats of the nut or bolt head in inches. For example, a ½ A/F spanner fits a nut ½ in. across the flats.

BSW and BSF spanners British Standard Whitworth and British Standard Fine nuts and bolts are no longer used on farm tractors and machinery, but they may still be found on older equipment. Spanners for BSW and BSF nuts and bolts are marked with the bolt diameter, but the distance between the jaw is much greater than this. For example, a spanner marked ½ in BSW has a jaw width of 15/16 in. The same spanner also fits 9/16 in. BSF nuts and bolts.

Unified National threads, both coarse and fine (UNC and UNF), were used before metric thread types came into common use in the UK. Bolts with these threads are usually marked with a line of small joined circles (see figure 22.9). They and the associated nuts can be adjusted using A/F spanners.).

Screwdrivers

There are several types of screwdriver with long, medium or short blade lengths and various blade widths for screws and bolts with slotted heads, together with a similar range of screwdrivers for screws with various types of cross-point head. Screwdriver handles may be made of wood or tough plastic, and some have insulated handles for electrical work. A screwdriver is for turning screws and small bolts; the blade will soon be damaged if it is used as a lever, and paint stirring is more effective with a stick!

Engineers' screwdrivers have a square or round metal blade with a tough plastic handle, which can, if necessary, be given a sharp tap to free a rusted screw. Another type of engineers' screwdriver has the metal blade extended through the handle, and if necessary it can be hit with a hammer.

Electricians' screwdrivers have tough insulated plastic handles, and many also have insulation on most of the length of the blade. They are meant for use on vehicle electrics and mains electrical equipment after disconnecting the supply.

Cabinet makers' screwdrivers have wooden handles. They are used for carpentry, and the handle should not be hit with a hammer, as such mis-treatment will eventually split the handle.

Cross-point screwdrivers are used for screws and small bolts with a cross-shaped slot. There are several types of cross-point screwdriver. Phillips screwdrivers have a pointed tip blade, while the end of a Posidriv® has a blunt point. A typical set of cross-point screwdrivers has five different point sizes.

Torx® screwdrivers have six external splines at the end of the blade which fit matching sized female splines in the screw or bolt head. Six different sizes are in common use. The different types of screw head are illustrated on page 287.

Ratchet handle screwdrivers, which are useful for high-speed work, can be set to undo or tighten screws and small bolts. Most ratchet handle screwdrivers have a magnetic bit holder at the end of the blade with set of different size slotted and cross-point bits stored in the handle.

Electric screwdrivers Some hand-held cordless electric drills with a reversible keyless chuck and adjustable torque settings can also be used as screwdrivers with slotted and cross head bits.

Cold Chisels

These are used to cut small pieces of metal, to split a nut that has seized on a bolt and for other metal work. Cold chisels in the farm workshop will have a cutting edge of between 6 to 25 mm wide and a handle between 100 mm to 250 mm long. A cold chisel has two taper angles, and the cutting edge should be sharpened with a bench grinder to an angle of about 60 degrees (see Figure 22.5). The cutting edge will be damaged if it is overheated and should be frequently cooled in cold water.

The end of a cold chisel handle will become mushroom-shaped with prolonged use. In this condition it is dangerous, as a glancing hammer blow may cause small pieces of metal to fly off the chisel handle, or the hammer may slip off the end of the chisel, sometimes with painful results. For safety's sake, grind off any burrs that form on the handle and wear goggles when chiselling metal or masonry.

A brick bolster with a much wider blade than a cold chisel, used for cutting bricks or concrete, is a useful addition to the tool kit. Most brick

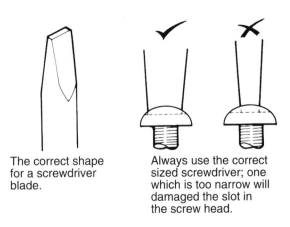

The correct shape for a screwdriver blade.

Always use the correct sized screwdriver; one which is too narrow will damaged the slot in the screw head.

Figure 22.4   Correct and incorrect sized screwdrivers.

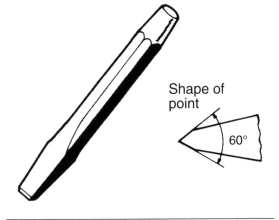

Shape of point

60°

Figure 22.5   Cold chisel.

bolsters have blade widths of 75 mm or 100 mm, and the best type of brick bolster has a hand guard on the handle.

Pliers

Engineers' or combination pliers are a necessary item in every tool kit. The plier jaws have flat faces for holding flat material and a rounded section for gripping small bolts and other round items. The inner section of the jaws has knife edges for cutting wire. Some engineers' pliers also have slots on the outside of the handles, close to the hinge, which can be used to cut wire.

Side cutting pliers are used to cut wire, split pins and similar items. They are also very useful for removing stubborn split pins when repairing a machine.

Long nose pliers have rather limited use, but are very useful for holding small parts in confined spaces. They should be used carefully, as rough use may break off the tips of the jaws.

Circlip pliers work in the opposite way to other pliers, with the handles being squeezed to open the jaws. They are used to remove circlips from shafts, bearings and other components and are made for both internal and external circlips.

Electricians' pliers may be of the side-cutting, engineers' or long nose type, all with heavily insulated handles. To avoid damage to the insulation, they should only be used for electrical work. Special pliers are made for stripping off short lengths of insulation when wiring an electric plug or an electrical component on a tractor. These pliers can be adjusted to suit various sizes of wire and provide an efficient way of stripping the insulation from it.

Files

Files are made in a variety of sizes, shapes and grades of cut. The teeth are on the body of the file, which has a tang at one end for attaching a wooden or plastic handle. The specification of a particular file depends on:

- *Length* The size of the cutting area, measured from the base to the tang at the end of the file.

- *Type of cut* Single-cut files have single rows of teeth running diagonally across the file, in one direction only. They are more suitable for filing hard steel. A single-cut file with a handle (Figure 22.7c) is generally known as a reaper file.

Double-cut files have teeth running diagonally across the face of the file in two directions so that the teeth are shaped like tiny pyramids. Double-cut files are used for most general filing work.

Rasps are very coarse files with individual teeth. They are used for cutting wood and very soft metals.

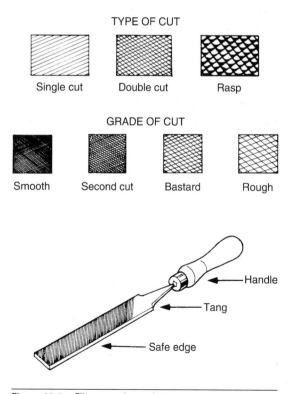

Figure 22.6   Files – grades and cuts.

Grade of cut This relates to the coarseness or fineness of the teeth.

Grades of file cut			
Type	**Teeth per cm**	**Teeth per inch**	**Use**
Dead smooth	27–43	70–110	Gives a polished precision finish.
Smooth	23–26	60–65	General filing work.
Second cut	11–16	28–42	A general-purpose file; it will remove metal quite quickly but does not give a smooth finish.
Bastard	8–12	22–32	Removes a lot of metal quickly but leaves a rough surface.
Rough	5–8	14–22	Makes a very coarse cut; a rasp.

The most useful files for the farm workshop are second cut. Bastard files will be needed for heavy filing, and some smooth-cut files for obtaining a good finish.

Types of file

Flat files, used for general work, are tapered in length and width. They have double-cut teeth on both faces and single-cut teeth on the edges.

Hand files are parallel along their length and have double-cut teeth. Some are safe edge files with single-cut teeth on only one edge. Safe edge files are used to prevent undercutting when filing into an internal corner.

Reaper files have single-cut teeth. They often have a shaped metal handle and are used for sharpening cutting edges.

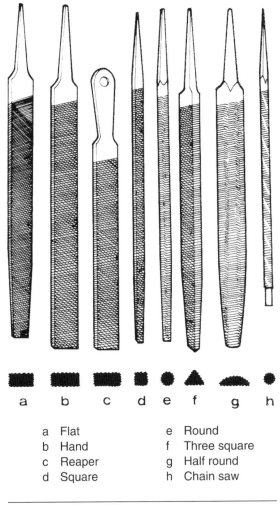

a	Flat	e	Round
b	Hand	f	Three square
c	Reaper	g	Half round
d	Square	h	Chain saw

Figure 22.7   Types of file.

Square files have double-cut teeth and are tapered towards the end. They are useful for filing out slots and square holes.

Round files also taper towards the end and are used to file out holes and round off internal corners. The larger files usually have second cut teeth and the shorter files are single cut.

Three square files are triangular in section and have double-cut teeth. Uses include filing angles of 60–90 degrees and working in awkward places where a flat file cannot be used.

Half round files have one flat face and one curved face. They are made with various combinations of teeth and the curved face is useful for forming a radius on a piece of metal.

Chain saw files are small diameter round files specifically made for sharpening a chain saw.

Using files

File teeth are designed to cut on the forward stroke; hand pressure applied on the cutting stroke should be released during the return stroke. Use the full length of the file and do not work too fast, one cutting stroke per second is best.

The teeth will become clogged if the file is used on dirty or rusty metal. They should be cleaned with a file card, which is similar to a wire brush but with short, stiff bristles on a flexible backing.

Do not store files so that they can rub against each other, as this will ruin the teeth. Wrap files separately in dry clean cloth or keep them upright in a stand. Always put a handle on the tang before using a file as the sharp point can cause a nasty injury to your hand.

Hammers

The best quality hammers are made of high carbon steel with a hickory wood or steel handle. Less expensive hammers may have an ash handle. The head is secured to the handle with hardwood and steel wedges.

Hammers are made in various sizes, denoted by the weight of the head. A small 0.25 kg (½ lb) hammer is useful for cutting paper gaskets and other light work. A 1.1 kg (2½ lb) is a basic item in any farm workshop tool kit, and plenty of work will be found for a 3.2 or 4.5 kg (7 or 10 lb) sledge hammer.

Engineers' or ball pein hammers are the most useful types of hammer. They have a ball at one end and a flat face at the other. The ball end is used for riveting.

Cross pein hammers have a blunt chisel-shaped head instead of a ball. They are mainly used for carpentry work.

A claw hammer with a nail drawing claw is another carpenter's tool, but one of these with a good quality steel shaft is useful in the farm workshop for any jobs involving the use of nails.

Soft faced hammers, usually with a copper face and a hide face, should be used when a steel faced hammer would damage the work. Examples include hammering soft or brittle materials such as brass and copper, and knocking out a tight bolt to avoid damaging the thread.

Club or lump hammers are bricklayers' tools, but their short handle and heavy head also makes them a handy general-purpose hammer.

A sledge hammer is a must on every farm. It is ideal for straightening bent components and driving steel posts into the ground. A large wooden mallet or beetle should be used for wooden stakes.

Using Hammers

Always hold a hammer close to the end of the handle and use your wrist as a pivot point. When the hammer makes contact with the work, the face should be parallel with the part being hammered.

Do not use a hammer with a loose or damaged handle. It is usually possible to tighten the head by driving an extra wedge into the head end of the handle. Hammer faces are very hard; do not strike two faces together as this may crack or damage them.

Hacksaws

Hacksaws for cutting metal usually have an adjustable frame suitable for blade lengths of 250 or 300 mm (10 or 12 in.).

The blade must be fitted with the teeth pointing away from the handle. It is tensioned with a wing nut on the opposite end of the frame from the handle. After taking up the slack, tighten the wing nut a further two or three turns. The blade should have the correct number of teeth

PLIERS

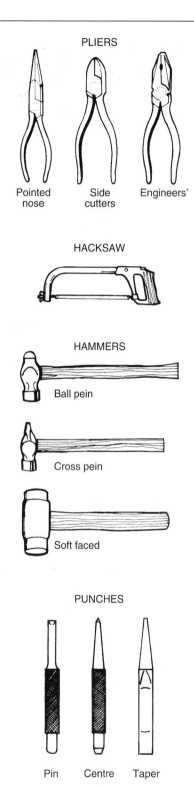

Pointed   Side   Engineers'
nose   cutters

HACKSAW

HAMMERS

Ball pein

Cross pein

Soft faced

PUNCHES

Pin   Centre   Taper

Figure 22.8   A selection of hand tools.

for the job in hand, and at least two teeth must be in contact with the metal. There is a choice of flexible or hard blades. A flexible blade is best for the handyman, as with careless use it may bend, but is unlikely to break. A hard blade gives a more accurate cut, but will snap if the blade is twisted on the cutting stroke.

- A coarse blade with 18 teeth per inch is best for thick sections and soft metals.
- For general work, use a blade with 24 teeth per inch.
- A fine blade with 32 teeth per inch is required for cutting very thin material.

Using Hacksaws

Always use the full length of the blade, making about one cutting stroke per second. Apply steady pressure on the saw during the forward cutting stroke and release that pressure on the return. Very little pressure is needed when cutting thin section materials. When possible, secure the work in a vice before starting to saw. Slacken the tension on the blade when the saw is not being used.

Punches

Various sizes and types of punch, usually with a knurled handle and a tapered or parallel pin, are another essential workshop tool. They can be used for driving pins, bolts or rivets out of holes, but the taper is likely to become stuck in the hole when attempting to drive out a long pin or bolt.

Parallel or pin punches are the same diameter throughout their length, making them ideal for driving out pins and bolts. They are made of hard steel, and the smaller sizes are likely to break if the punch is used as a lever.

Centre punches are used to mark the centre spot when drilling a hole in metal. The impression made by the punch helps to start the drill bit. A centre punch should be sharpened on a grindstone so that the point angle is about 50 degrees.

WORKSHOP POWER TOOLS

Drills

An electric hand drill is an essential workshop power tool. There are many types of cordless and mains electric drill, including small domestic models suitable for drilling holes up to 8 mm in diameter, and typical farm workshop models for drilling up to 13 mm holes. Larger drills, attached to a bench drill stand, can, depending on size, be used to drill holes of 19 mm diameter in steel; and holes of up to 25 mm or more can be drilled with a free-standing pillar drill with a 1 hp electric motor.

Compressed air-driven drills for use with a portable compressor or permanent airline circuits in the workshop provide an alternative, and their use eliminates the risk of an electric shock.

Mains electric drills may be plugged in to a 13 amp, 240 volt socket or used with a transformer at 110 volts. Cordless drills, with a

Plate 22.2   This 110/240 volt variable speed hammer drill with a forward and reverse control has a Jacobs chuck and a depth gauge. (Makita)

clip-on rechargeable 14.4, 18 or 36 volt battery, are supplied with their own battery charger. Most drills can be switched from high to low speed. The variable two-speed drill illustrated in Plate 22.2 has a maximum speed of 2,900 rpm in the high setting and up to 1,200 rpm in low. Many mains electric and cordless drills have a hammer setting, which is useful for drilling holes in masonry. It is important to use the correct speed for the job in hand. It is better to use the slower speed when drilling holes over 5 mm in steel and other hard materials.

Some drills have a chuck to grip the drill bit with a key to tighten the twist drill in the chuck, but many have a keyless chuck. The body of the chuck is pushed upward against a spring to accept the twist drill and releasing the pressure locks it in the chuck jaws. Many cordless drills can also be used as electric screwdrivers.

Twist Drills

There are various types of high- and low-speed twist drills for metal, wood, masonry and other materials. The cheaper, low-speed twist drills, made of carbon steel, have a bright appearance but they are only suitable for hand drills. Twist drills made of high-speed steel, often black in colour, should be used to drill steel and other hard metals with an electric drill.

Plate 22.1   This cordless electric 18 volt hammer drill with variable forward/reverse speeds and range of torque settings can also be used as an electric screwdriver. (Makita)

Twist drills can be purchased singly or in sets. Metric twist drill sizes usually increase in steps of 0.5 mm for farm workshop use. A typical 25-piece high-speed metric set has twist drills ranging from 1 to 13 mm.

Using Drills

- Check that the twist drill is fully inserted into the chuck. For chucks with a key, tighten the twist drill using the key in all three positions.
- Use sharp twist drills. There is little point in trying to force a bit into metal if it is blunt.
- Clamp the work securely in a vice.
- A centre punch dot mark will help to start the drill bit into work.
- Large diameter holes are easier to drill if a small pilot hole is drilled first. Use a slow drill speed for holes over 5 mm diameter in steel and other hard materials.
- Care is needed when drilling holes in thin sheet metal. Clamp the sheet down so that it cannot pick up on the twist drill when the hole is almost complete. For the best results, it helps if a piece of wood is placed under the metal sheet before drilling the hole.

- The drill will cut better and give a good finish if it has suitable lubrication. Use water-soluble oil when drilling steel. This is a white fluid that can be bought from a tool shop and diluted with water. Cutting oil, as it is known, has good lubricating properties and helps to cool the twist drill. Cast iron should be drilled dry, and most soft metals are also drilled without lubrication.

Bench Grinders

An electric grinder is a useful power tool in the farm workshop. A floor-standing pedestal grinder will be needed for heavier work, but a bench grinder with 150 or 200 mm diameter abrasive wheels should be adequate for most needs in the farm workshop. Bench grinders will either have coarse and fine abrasive wheels or an abrasive wheel and a rotary wire brush. The fine abrasive wheel should be reserved for sharpening twist drills, cold chisels and hand tools such as bill hooks and axes.

Safety regulations require that any person fitting an abrasive wheel to a workshop grinder has been trained in the correct procedures for this task.

Plate 22.3   Bench grinders can be fitted with abrasive grinding wheels or wire brushes. This one has 150 mm diameter abrasive wheels and transparent spark guards. (Sealey)

Using Bench Grinders

- Always wear goggles to protect your eyes from flying sparks and stray chips of metal.
- Try to use the full face of the grinding wheel and do not put any pressure on the side of the wheel.
- Set the work rests close up against the grinding wheel; the metal being ground may be pulled between the rest and the wheel if the gap is too wide.
- Keep the wheel's circumference in good condition. It will become uneven with use and can be restored to the correct shape with a stone dresser. One type has a set of hard rotary cutters, which is held against the face of the running wheel until the edge is flat.

Angle Grinders

Angle grinders have a grinding or cutting disc running at right angles to the motor drive shaft. Sizes vary, from a lightweight 12 or 18 volt cordless angle grinder with a 100 mm diameter grinding disc and 75 mm cutting disc, to a heavy-duty mains electric model with 230 mm cutting and grinding discs. Sanding discs are made in various grades of abrasiveness, or grits, from fine to coarse.

Angle grinders are useful for cutting or grinding large pieces of metal and sharpening hedge cutter flails. Small angle grinders can also be used with a sanding disc or a wire brush.

Using Angle Grinders

- Always wear goggles. It is also advisable to wear a particle mask as protection against harmful dust from metal, masonry and from the disc as it wears away. Wear a pair of strong gloves when angle grinding and ear defenders to protect your hearing.
- Check that no other person is at risk from flying sparks when using an angle grinder.
- Clamp small parts to a bench before using an angle grinder and make sure the electric cable is in a safe position before starting work.

Compressors

Most farm workshop air compressors are driven by an electric motor. Some are portable units on wheels; others are permanent installations with a fixed airline around the workshop.

A portable compressor with a small air receiver or a permanent installation with large air receiver has many uses in the farm workshop, including inflating tyres, spray painting and cleaning down equipment. Compressed air can also be used to drive a range of power tools, including drills and angle grinders.

Plate 22.4   Angle grinders are used mainly for cutting through metals, and can be adapted for sanding various materials. (Makita)

Plate 22.5   Compressors have a variety of uses in the farm workshop. They include inflating tyres, paint spraying and driving air-operated power tools including drills, angle grinders and screwdrivers. This portable compressor for use with a 13 amp socket has a 50 litre air cylinder. (Sealey)

Using Compressors

- Wear goggles when using compressed air to blow dust and dirt from machinery and equipment.
- Never direct a jet of compressed air at a person's body as this can cause serious injury.
- Water is formed when air is compressed, and this will collect in the air receiver tank. It should be drained away at frequent intervals to prevent damage to the tank.

Battery Chargers

A small 4 amp capacity trickle charger, connected to an electric socket, will be sufficient to keep stored batteries in a satisfactory state of charge or pep up an elderly tractor battery if it is left on charge overnight.

Plate 22.6 This battery charger may be used for low rate trickle charging, or to boost a 12 or 24 volt battery for a short while before starting an engine with an almost flat battery. (Sealey)

A high rate battery charger can be used to charge two or more batteries at the same time or boost a discharged battery for a short time before starting the engine.

Welding Equipment

A welder is another essential piece of workshop equipment. The main types of electric welder are MMA (manual metal electric arc), MIG (metal inert gas), MAG (metal active gas) and TIG (tungsten inert gas).

Manual metal arc welding A manual metal arc welding set consists of an adjustable output mains-operated transformer, heavy-duty cables for the earth clamp and an electrode holder used to grip a flux-coated welding rod. When the welder is used, heat produced by an electric current (arc) is passed from the welding rod to the work piece. This melts the metal welding rod, and the edges of the two parts are

Plate 22.7 Electric arc and MIG or TIG welding can be done with this electric welding set. (Sealey)

joined together. The molten metal is deposited in a narrow gap that is left between the two parts to complete the joint. The flux also melts to create a gas shield to protect the joint from contaminants and forms a slag to slow the cooling rate of the joint. Small output DC welders suitable for welding thinner section metals can be run from a 13 amp socket.

MIG/MAG welding Metal Inert Gas and Metal Active Gas welders are semi-automatic, with a continuous flow of wire filler rod from a spool at a controlled rate to the electrode holder. Gas is used to create a protective shield around the electrode and prevent the formation of impurities in the molten metal as it is placed in the joint. MIG welding is faster than metal arc welding, and there is no slag over the joint, but a separate cylinder is required to provide a gas for the shield. Argon is used to provide the gas shield when welding aluminium. A gas shield of either carbon dioxide or a mixture of argon and carbon dioxide is used when welding steel.

TIG Welding Heat is created by an electric arc formed between the tip of a tungsten electrode in a hand-held torch and the metal being welded. For most welding jobs, a separate hand-held filler wire is used to complete the weld. A gas shield, usually of pure argon, protects the weld from contamination. A TIG welder is suitable for high-quality welding of steel, stainless and other alloy steels and aluminium.

WORKSHOP MATERIALS

All workshops have at least a small stock of metal, nuts, bolts, washers and other hardware items. Many will also have a selection of replacement parts such as filters, fan belts, shares and tines for ploughs and cultivators, and similar items. A variety of metals and other materials are used in the construction of agricultural machinery. Metals and alloys – a mixture of two or more metals – have various physical properties that make them ideal

for some applications and totally unsuitable for others. Cast iron, which is brittle, is ideal for making intricate castings but quite useless for a trailer drawbar.

A basic knowledge of the more important metals is needed by anyone using a workshop to make or mend large or small pieces of equipment.

Ferrous Metals

Ferrous metals contain iron. There are many types, including cast iron, carbon steel and alloy steels, used in the manufacture of farm machinery. Ferrous metals contain carbon in varying percentages according to the type of iron or steel.

Cast iron is relatively cheap and is easy to drill, machine and weld. It is rather brittle, but has great strength in compression (i.e. it resists crushing forces). It is used to make castings, as the name suggests, sometimes of quite intricate shapes. Molten cast iron is poured into moulds made from sand to produce the required shape of casting.

Most types of cast iron contain between 3 and 4 per cent carbon, plus small amounts of manganese, phosphorus, silicon and sulphur. The actual amount of these elements varies with the type of iron.

When molten iron is allowed to cool naturally in a casting box, grey cast iron is produced. Rapid chilling, usually with a chilling block in the casting box, results in white cast iron or chilled cast iron, which is much harder and more brittle.

Mild steel or low carbon steel is the cheapest and most common type of ferrous metal used in the manufacture of farm machinery. Mild steel contains from 0.1 to 0.3 per cent carbon. It is easy to saw, drill, machine, bend and weld, but it has no great strength and will soon rust if it is not protected from the weather. Painting, galvanising (coating with zinc) and plastic coating are some of the methods used to combat corrosion.

Uses include tractor bonnets, machinery guards, farm gates, etc. Mild steel is readily available from metal merchants in many shapes, including sheets and tubes, or in round, flat or square sections.

Medium carbon steel is harder and tougher than mild steel, and contains between 0.3 and 0.7 per cent carbon. It can be forged, welded, machined, drilled or sawn, but its increased hardness makes these processes more difficult. It can be annealed (softened) to make the machining processes easier, but will then need further heat treatment to restore its hardness. Medium carbon steel is used for shafts and axles, plough beams and discs, cultivator tines, etc.

High carbon steel has a carbon content ranging from 0.7 to 1.4 per cent. A higher carbon content indicates that it is a hard but rather brittle form of steel. It is used to make good quality tools including chisels, punches, files and twist drills.

Alloy Steel contains one or more non-ferrous metals to give it additional properties. Some examples of alloy steel include:

- *Stainless steel*, which contains chromium to give it non-corrosive properties.
- *Tool steel* has a small quantity of tungsten, which gives a good cutting edge.
- *Nickel steel* is a very tough metal and has the necessary strength for making high-quality steel axles and shafts.

Heat treatment of steel Various heat treatments are required when manufacturing tools and machinery components from carbon steel. Mild steel, because of its low carbon content, is not suitable for most heat treatment processes.

- *Hardening* Medium and high carbon steels can be hardened by heating the metal to red heat and then quenching it in cold water.

Quenching (rapid cooling) gives added hardness but makes the metal rather brittle.

- *Tempering* removes the brittleness from hardened steel. The metal is re-heated to a certain temperature and then quenched again. As the temperature of the metal increases, its colour changes from straw to red, brown, purple and blue. These colours are used to determine the quenching point.

  A cold chisel is tempered by applying heat to the handle above the cutting edge, which is first polished with emery cloth. The chisel is quenched when the cutting edge is dark purple.

- *Annealing* softens metal to make it easier to machine. Steel is annealed by heating it to red heat and allowing it to cool very slowly.

- *Case hardening* This is a factory process of heating and quenching to give steel a hard-wearing surface with a softer, shock-resistant centre.

Non-Ferrous Metals

There are many different non-ferrous metals and alloys, but very few of them are used in the manufacture of farm machinery. The more important non-ferrous metals and alloys include:

Copper A good conductor of heat and electricity, copper can be welded, is easy to work with and resists corrosion. It is a soft metal, but has the strange property of work hardening. This means that it will become harder with age or when it is consistently hammered or bent. Copper can be softened by annealing. It is heated to a cherry red colour and then quenched in clean cold water.

Zinc is a corrosion-resistant metal used to coat steel sheets, tanks and pipes. The process of coating steel with zinc is called galvanising.

Aluminium is a very light material of moderate strength. In its pure form, aluminium is sometimes used to make sheeting. When alloyed with other metals including copper, zinc and silicon, it is used to make castings including pistons, cylinder blocks and gear cases for small engines.

Brass is an alloy of copper and zinc. There are two main types of brass. Soft brass, which contains about 40 per cent zinc, is easy to work and is produced as sheets, tubes, wire and other items. Hard brass contains more than 40 per cent zinc and is used to make castings.

Bronze, of which there are several types, is basically an alloy of copper and tin with small amounts of other metals added to give it special properties. Phosphorus is added to make phosphor bronze, which is a good bearing material.

Identification of Metals

The basic non-ferrous metals are easily recognised by their colour but identifying the different types of iron and steel requires more knowledge.

The hardness of metal can be determined with a hacksaw or file. Mild steel, grey cast iron and most ferrous metals are relatively easy to cut. Medium and high carbon steels are much more difficult. The grained structure of cast iron also distinguishes it from steel. Another test to identify the different types of iron and steel can be done with an electric grinder. The sparks from the different metals are a good aid to their identification.

- *Mild steel* produces large quantities of white spear-shaped sparks and a few stars.
- *Medium carbon steel* gives off fewer of these white sparks with a lot of star-shaped sparks mixed in with them.
- *High carbon steel* produces fewer white star-shaped stars.
- *Cast iron* gives a small quantity of reddish sparks which turn to a straw colour.

Plastics

Plastic does not corrode, and for this reason it is widely used for many agricultural applications, especially for fertiliser spreaders and crop sprayers. There are two basic types of plastic:

Polythene and PVC are two examples of thermoplastic plastic. These materials will become soft and pliable if they are subjected to gentle heat, but excessive heat will result in permanent distortion.

Thermosetting plastic is hardened during the manufacturing process and will not soften or melt when subjected to gentle heat. However, like all plastics, it will be destroyed if exposed to high temperatures.

Nuts and bolts

Many different types of nut and bolt will be found on farm machinery and equipment.

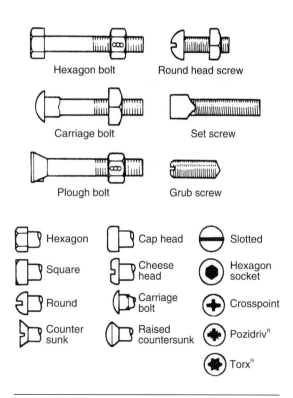

Figure 22.9   Types of nut and bolt.

Hexagon-headed bolts require no description.

- **Carriage bolts** are used to fix timber to timber or timber to steel. The square section under the bolt head pulls into the wood to stop the bolt turning while tightening the nut.
- **Plough bolts** are made in various patterns, and are used, for example, to fix a plough mouldboard to the frog. Its countersunk head stops the bolt turning in the hole, and the flat top gives a smooth surface on the face of the mouldboard.

Machine screws are smaller in diameter than bolts. They have round or countersunk heads, are threaded right up to the head of the bolt and may have hexagonal or square nuts.

Set screws, often with a square head, are used in threaded holes to lock gears, pulleys or other components onto a shaft.

Grub screws serve a similar purpose, but they have screwdriver slots, and when tightened they are flush with the pulley or gear hub.

Most nuts are hexagon-shaped, but some square nuts are used. Wing nuts can be tightened with the fingers and are useful where frequent adjustments are necessary. There are several types of lock nut. Some have a soft metal or plastic insert at the top of the nut, with the hole in the insert slightly smaller than the bolt thread. When the nut is tightened, the thread cuts into the insert, and this is sufficient to lock the nut in position. Another type of lock nut is castle-shaped, with a split pin which is passed through the bolt after it has been lined up with two of the slots in the castle nut. When machine parts, such as a lever, need to pivot on a bolt, two nuts with one tightened against the other can be used.

The tendency for nuts to vibrate loose can be overcome by putting a spring washer or a star washer under the nut. Sharp edges on the washer grip the nut and keep it tight on the bolt. A tab washer (a flat washer with two or more locking tabs) can also be used. One of the tabs is hammered up against the nut, and the other is bent down over the edge of the component being secured by the nut.

6TH EDITION FARM MACHINERY – INDEX